# CAMBRIDGE LIBRARY COLLECTION

*Books of enduring scholarly value*

## History

The books reissued in this series include accounts of historical events and movements by eye-witnesses and contemporaries, as well as landmark studies that assembled significant source materials or developed new historiographical methods. The series includes work in social, political and military history on a wide range of periods and regions, giving modern scholars ready access to influential publications of the past.

## The Wonderful Century

The British naturalist and explorer Alfred Russel Wallace (1823–1913) was one of the leading evolutionary thinkers of the nineteenth century. He is best known for working in parallel with Darwin on evolution and natural selection. A social reformer and a prolific writer, he criticised the social and economic system in nineteenth-century Britain, and raised concerns over the environmental impact of human activity. First published in 1898, this book looks back over the history of the nineteenth century, and describes its material and intellectual achievements with the aim 'to show how fundamental is the change they have effected in our life and civilization'. The book surveys technological inventions such as the railway, the telegraph and telephone, as well as photography. But it also analyses the century's 'failures', and discusses the issues of poverty, greed and militarism.

Cambridge University Press has long been a pioneer in the reissuing of out-of-print titles from its own backlist, producing digital reprints of books that are still sought after by scholars and students but could not be reprinted economically using traditional technology. The Cambridge Library Collection extends this activity to a wider range of books which are still of importance to researchers and professionals, either for the source material they contain, or as landmarks in the history of their academic discipline.

Drawing from the world-renowned collections in the Cambridge University Library, and guided by the advice of experts in each subject area, Cambridge University Press is using state-of-the-art scanning machines in its own Printing House to capture the content of each book selected for inclusion. The files are processed to give a consistently clear, crisp image, and the books finished to the high quality standard for which the Press is recognised around the world. The latest print-on-demand technology ensures that the books will remain available indefinitely, and that orders for single or multiple copies can quickly be supplied.

The Cambridge Library Collection will bring back to life books of enduring scholarly value (including out-of-copyright works originally issued by other publishers) across a wide range of disciplines in the humanities and social sciences and in science and technology.

# The Wonderful Century

## *Its Successes and its Failures*

Alfred Russel Wallace

Cambridge University Press

CAMBRIDGE UNIVERSITY PRESS

Cambridge, New York, Melbourne, Madrid, Cape Town,
Singapore, São Paolo, Delhi, Tokyo, Mexico City

Published in the United States of America by Cambridge University Press, New York

www.cambridge.org
Information on this title: www.cambridge.org/9781108036795

© in this compilation Cambridge University Press 2011

This edition first published 1898
This digitally printed version 2011

ISBN 978-1-108-03679-5 Paperback

This book reproduces the text of the original edition. The content and language reflect the beliefs, practices and terminology of their time, and have not been updated.

Cambridge University Press wishes to make clear that the book, unless originally published by Cambridge, is not being republished by, in association or collaboration with, or with the endorsement or approval of, the original publisher or its successors in title.

THE WONDERFUL CENTURY

Every young soul, ardent and high, rushing forth into life's hot fight;
Every home of happy content, lit by love's own mystical light;
Every worker who works till the evening, and earns before night his wage,
Be his work a furrow straight drawn, or the joy of a bettered age;
Every thinker who, standing aloof from the throng, finds a high delight
In striking with tongue or with pen a stroke for the triumph of right,—
All these know that life is sweet; all these, with a consonant voice,
Read the legend of Time with a smile, and that which they read is "Rejoice!"

—*Sir Lewis Morris.*

# THE WONDERFUL CENTURY

## ITS SUCCESSES AND ITS FAILURES


BY

ALFRED RUSSEL WALLACE

*Author of "The Malay Archipelago," "Darwinism," etc., etc*

London

SWAN SONNENSCHEIN & CO., LIMD

NEW YORK: DODD, MEAD & CO

1898

The old times are dead and gone and rotten;
  The old thoughts shall never more be thought;
The old faiths have failed and are forgotten,
  The old strifes are done, the fight is fought;
And with a clang and roll, the new creation
Bursts forth, 'mid tears and blood and tribulation.

*—Sir Lewis Morris.*

*Copyright* 1898, (*in the United States of America*
*by* DODD, MEAD & CO., *New York.*

# PREFACE

The present work is not in any sense a history, even on the most limited scale. It may perhaps be termed an appreciation of the Century—of what it has done, and what it has left undone. The attempt has been made to give short, descriptive sketches of those great material and intellectual achievements which especially distinguish the Nineteenth Century from any and all of its predecessors, and to show how fundamental is the change they have effected in our life and civilization. A comparative estimate of the number and importance of these achievements leads to the conclusion that not only is our century superior to any that have gone before it, but that it may be best compared with the whole preceding historical period. It must therefore be held to constitute the beginning of a new era of human progress.

But this is only one side of the shield. Along with these marvellous Successes—perhaps in consequence of them—there have been equally striking Failures, some intellectual, but for the most part moral and social. No impartial appreciation of the century can omit a reference to them; and it is not improbable that, to the historian of the future, they will be considered to be its most striking characteristic. I have therefore given them due prominence. No doubt it will be objected that I have devoted far too much space to them—more than half the volume. But this was inevitable, for the very obvious reason that, whereas the successes are universally admitted and had only to be *described*, the failures are either ignored or denied, and therefore required to be *proved.* It was thus necessary to give a tolerably full summary of the evidence in every case in which an allegation of failure has been made. The Vaccination question has been discussed at the greatest length for several reasons. It is the only surgical operation that, in our country, has ever been universally enforced by law. It has been recently inquired into by a Royal Commission, whose Majority Report is directly opposed to the real teaching of the official and national statistics presented in the detailed reports. The operation is, admittedly, the cause of many deaths, and of a large but unknown amount of permanent injury; the only really trustworthy statistics on a large scale prove it to be wholly without effect as a preventive of small-pox; many hundreds of persons are annually punished for refusing to have their children vaccinated; and it will undoubtedly rank as the greatest and most pernicious failure of the century. I claim, that the evidence set forth in this chapter, with the diagrams which illustrate it, demonstrate this conclusion. It is no longer a question of opinion, but of science; and I have the most complete confidence that the result I have arrived at is a statistical, and therefore a mathematical certainty.

Of even greater importance, though less special to the century, is the perennial problem of wealth and poverty. In dealing with this question I have adduced a body of evidence showing that, accompanying our enormous increase of wealth, there has been a corresponding increase of poverty, of insanity, of suicide, and probably even of crime, together with other indications of moral and physical deterioration. To the facts I have set forth I earnestly call the attention of all those who have at heart the progress of true civilization and the welfare of humanity.

A. R. W.

Parkstone, Dorset,
*April*, 1898.

If thou would'st make thy thought, O man, the home
Where other minds may habit, build it large.
Make its vast roof translucent to the skies,
And let the upper glory dawn thereon,
Till morn and evening, circling round, shall drop
Their jewelled plumes of sun-flame and of stars.

—*Thomas Lake Harris.*

# CONTENTS

## *PART I.—SUCCESSES*

## 1898.

And these are ours to-day! The boundless flood
  Of infinite Research—the ocean vast
Of endless exploration—and our barque
  Of Science builded, fairly launched at last,
Captain'd by Thought—by Reason piloted—
  Sails forth upon the venture—and to us,
To search the shores of Doubt—in midnight hid;
  To give, if such there be, new Worlds to light,
  And that we have, with better day make bright.

—*J. H. Dell.*

# Part I—Successes

## CHAPTER I

### MODES OF TRAVELLING

Put forth your force, my iron horse, with limbs that never tire!
The best of oil shall feed your joints, and the best of coal your fire;
Like a train of ghosts, the telegraph posts go wildly trooping by,
While one by one the milestones run, and off behind us fly!
Dash along, crash along, sixty miles an hour!
Right through old England flee!
For I am bound to see my love,
Far away in the North Countrie.

—*Prof. Rankine.*

WE men of the Nineteenth Century have not been slow to praise it. The wise and the foolish, the learned and the unlearned, the poet and the pressman, the rich and the poor, alike swell the chorus of admiration for the marvellous inventions and discoveries of our own age, and especially for those innumerable applications of science which now form part of our daily life, and which remind us every hour of our immense superiority over our comparatively ignorant forefathers.

But though in this respect (and in many others) we undoubtedly think very well of ourselves, yet, in the opinion of the present writer, our self-admiration does not rest upon an adequate appreciation of the facts, No one, so far as I am aware, has yet pointed out the altogether exceptional character of our advance in science and the arts, during the century which is now so near its close. In order to estimate its full importance and grandeur—more especially as regards man's increased power over nature, and the application of that power to the needs of his life to-day, with unlimited possibilities in the future—we must compare it, not with any preceding century, or even with the last millennium, but with the whole historical period—perhaps even with the whole period that has elapsed since the stone age.

Looking back through the long dark vista of human history, the one step in material progress that seems to be really comparable in importance with several of the steps we have just made, was, when Fire was first utilised, and became the servant and the friend, instead of being the master and the enemy of man. From that far distant epoch even down to our day, fire, in various forms and in ever-widening spheres of action, has not only ministered to the necessities and the enjoyments of man, but has been the greatest, the essential factor, in that continuous increase of his power over nature, which has undoubtedly been a chief means of the development of his intellect and a necessary condition of what we term civilisation. Without fire there would have been neither a bronze nor an iron age, and without these there could have been no effective tools or weapons, with all the long succession of mechanical discoveries and refinements that depended upon them. Without fire there could be no rudiment even of chemistry, and all that has arisen out of it. Without fire much of the earth's surface would be uninhabitable by man, and much of what is now wholesome food would be useless to him. Without fire he must always have remained ignorant of the larger part of the world of matter and of its mysterious forces. He might have lived in the warmer parts of the earth in a savage or even in a partially civilised condition, but he could never have risen to the full dignity of intellectual man, the interpreter and master of the forces of nature.

Having thus briefly indicated our standpoint, let us proceed to sketch in outline those great advances in science and the arts which are the glory of our century. In the course of our survey we shall find, that the more important of these are not mere improvements upon, or developments of, anything that had been done before, but that they are entirely new departures, arising out of our increasing knowledge of and command over the forces of the universe. Many of these advances have already led to developments of the most startling kind, giving us such marvellous powers, and such extensions of our normal senses, as would have been incredible, and almost unthinkable even to our greatest men of science, a hundred years ago. We begin with the simplest of these advances, those which have given us increased facilities for locomotion.

The younger generation, which has grown up in the era of railways and of ocean-going steamships, hardly realise the vast change which we elders have seen, or how great and fundamental that change is Even in my own boyhood the wagon for the poor, the stage-coach for the middle class, and the post-chaise for the wealthy, were the universal means of communication, there being only two short railways then in existence—the Stockton and Darlington opened in 1825, and the Liverpool and Manchester line opened in 1830. The yellow post-chaise, without any driving seat, but with a postilion dressed like a jockey riding one of the pair of horses, was among the commonest sights on our main roads; and together with the hundreds of four-horse mail and stage coaches, the guards carrying horns or bugles which were played while passing through every town or village, gave a stir and liveliness and picturesqueness to rural life which is now almost forgotten.

When I first went to London (I think about 1835) there was still not a mile of railroad in England, except the two above-named, and none between London and any of our great northern or western cities were even seriously contemplated. The sites of most of our great London railway termini were then on the very outskirts of the suburbs; Chalk Farm was a genuine farmhouse, and Primrose Hill was surrounded by open fields.

A few years later (in 1837–38) I was living near Leighton Buzzard while the London and Birmingham Railway, the precursor of the present London and North-Western system, was in process of construction; and when the first section was opened to Watford, I travelled by it to London, third-class, in what is now an ordinary goods truck, with neither roof nor seats, nor any other accommodation than is now given to coal, iron, and miscellaneous goods. If it rained, or the wind was cold, the passengers sat on the floor and protected themselves as they could. Second-class carriages were then what the very worst of the third-class are or were a few years ago—closed in, but low and nearly dark, with plain, wooden seats—while the first class were exactly like the bodies of three stage coaches joined together. The open passenger trucks were the cause of much misery, and a few deaths from exposure, before they were somewhat improved; but even then there was evidently a dread of making them too comfortable, so a roof was put to them, also seats, and the sides a little raised but open at the top, about equal in comfort to our present cattle trucks. At last, after a good many years, the despised third-class passengers were actually provided with carriages of the early second-class type; and it is only in comparatively recent times that the greater railway companies realised the fact that third-class passengers were so numerous as to be more profitable than the other two combined, and that it was worth while to give them the same comfort, if not the same luxury, as those who could afford to travel more expensively.

The continuous progress in speed and comfort is matter of common knowledge, and nothing more need be said of it here. The essential point for our consideration is, the fundamental and even revolutionary nature of the change that has been wholly effected during the present century. In all previous ages the only modes of travelling or of conveying goods for long distances were by employing either men or animals as the carriers. Wherever the latter were not used all loads had to be carried by men, as is still the case over a large part of Africa, and as was the case over almost the whole of America before its discovery by the Spaniards.

But throughout Europe and Asia the horse was domesticated in very early times, and was used for riding and in drawing war chariots; and throughout the Middle Ages pack-horses were in universal use for carrying various kinds of goods and produce, and saddle horses for riding. All journeys were then made on horseback, and it was in comparatively recent times that wheeled vehicles for travelling in came into general use in England. The very first carriage was made for Queen Elizabeth in 1568; the first that plied for hire in London were in 1625, and the first stage coaches in 1659.

But chariots drawn by horses were used, both in war and peace, by all the early civilized peoples. Pharaoh made Joseph ride in a chariot, and he sent wagons to bring Jacob, with his children and household goods, to Egypt. A little later chariots were sent by the Syrians as tribute to Pharaoh. Homer describes Telemachus as travelling from Pylos to Sparta in a chariot provided for him by Nestor,—

> "The rage of thirst and hunger now suppress'd,
> The monarch turns him to his royal guest;
> And for the promis'd journey bids prepare
> The smooth-haired horses, and the rapid car."

It is clear, therefore, that in the earliest historic times all the various types of wheeled vehicles were used—for war, for racing, for travelling, and for the conveyance of merchandise. They must also have been used throughout a large part of Europe, since Cæsar found our British ancestors possessed of war-chariots, which they managed with great skill, implying a long previous acquaintance with the domesticated horse and its use in humbler wheeled vehicles.

Thus, throughout all past history the modes of travelling were essentially the same, and an ancient Greek or Roman, Egyptian or Assyrian, could travel as quickly and as conveniently as could Englishmen down to the latter part of the eighteenth century. It was mainly a question of roads, and till the beginning of the nineteenth century our roads were for the most part far inferior to those of the Romans. It is, therefore, not improbable that during the Roman occupation of Britain the journey from London to York could have been made actually quicker than a hundred and fifty years ago.

We see, then, that from the earliest historic, and even in prehistoric times, till the construction of our great railways in the second quarter of the present century, there had been absolutely no change in the methods of human locomotion; and the speed for long distances must have been limited to ten or twelve miles an hour even under the most favourable conditions, while generally it must have been very much less. But the railroad and steam-locomotive, in less than fifty years, not only raised the speed to fifty or sixty miles an hour, but rendered it possible to carry many hundreds of passengers at once with punctuality and safety for enormous distances, and with hardly any exposure or fatigue. For the civilised world, travelling and the conveyance of goods have been revolutionized, and by means which were probably neither anticipated nor even imagined fifty years before.

Dr. Erasmus Darwin, who predicted steam carriages, had apparently no conception of the possibility of railroads, the enormous cost of which would have seemed to be prohibitory. And we have by no means yet fully developed their possibilities, since even now a railroad could be made on which we might safely travel more than a hundred miles an hour, it being merely a question of expense.

In steam navigation there has been a very similar course of events, with the same characteristic of a completely new departure, leading to unknown developments and possibilities. From the earliest dawn of history men used rowing or sailing vessels for coasting trade or for crossing narrow seas. The Carthaginians sailed nearly to the equator on the west coast of Africa, and in the eleventh century the Northmen reached North America on the coast of New England. Exactly five hundred years ago Vasco da Gama sailed from Portugal round the Cape of Good Hope to India, and in the next century Columbus and his Spanish followers crossed the Atlantic in its widest part to the West Indies and Mexico. From that time sailing ships were gradually improved, till they culminated in our magnificent frigates for war purposes and the clipper ships in the China and Australian trade, which were in use up to the middle of the century. But during all this long course of development there was no change whatever in principle, and the grandest three-decker or full-rigged clipper ship was but a direct growth, by means of an infinity of small modifications and improvements, from the rudest sailing boat of the primeval savage.

Then, at the very commencement of the present century, the totally new principle of steam-propulsion began to be used, at first experimentally and with many failures, on rivers, canals, and lakes, till about the year 1815 coasting steamships of small size came into pretty general use. These were rapidly improved; but it was not till the year 1838 that the *Great Western*, of 1,340 tons and 400 horse-power, made the passage from Bristol to New York in fourteen days, and thus inaugurated the system of ocean steam-navigation, which has since developed to such an enormous extent. The average speed then attained, of about ten miles an hour, has now been more than doubled, and is still increasing. But the horse-power, needed to attain this high speed, has increased in much greater proportion; and it is only the much greater size and capacity, both for passengers and goods, that render such high speeds and enormous consumption of coal profitable. Some of the smaller steel-built war-ships — torpedo-boats and torpedo-destroyers — have considerably exceeded thirty miles an hour, and the limit of speed is probably not yet reached. Many suggested forms of vessel, such as the cigar-shaped and the roller-boats, have not been adequately tried; and there are other suggested forms by means of which greater steadiness and speed may yet be obtained.

Almost as remarkable as our railroads and steamships is the new method of locomotion by means of the bicycle and tricycle. The principle is old enough, but the perfection to which these vehicles have now attained has been rendered possible, by the continuous growth of all kinds of delicate tools and machines required in the construction of the infinitely varied forms of steam-engines, dynamos, and other rapidly-moving machinery. In the last century it would not have been possible to construct a modern first-class bicycle, even if any genius had invented it, except at a cost of several hundred pounds. The combination of strength, accuracy, and lightness would not then have been attainable. It is a very interesting fact that three out of the four methods of rapid locomotion we now possess should have attained about the same maximum speed. The racehorse, the steamship, and the bicycle, have each of them reached thirty miles an hour. The horse is, however, close upon, if it has not actually attained, its utmost limits;

the bicycle can already beat the horse for long distances, and will certainly go at higher speeds for short ones; while the steamship will also go much quicker, though how much no one can yet say. The greatest possibilities are with the bicycle, driven by electric power or compressed air, by which means, on a nearly straight and fairly level, asphalt track, no doubt fifty miles an hour will soon be reached.

We see, then, that during the nineteenth century three distinct modes of locomotion have been originated and brought to a high degree of perfection. Two of them, the locomotive and the steamship, are altogether different in principle from what had gone before. Up to the very times of men now living, all our locomotion was on the same old lines which had been used for thousands of years. It had been improved in details, but without any alteration of principle and without any very great increase of efficiency. The principles on which our present methods rest are new; they already far surpass anything that could be effected by the older methods; with wonderful rapidity they have spread over the whole world, and they have in many ways modified the habits and even the modes of speech of all civilised peoples.

This vast change in the methods of human locomotion, already so ubiquitous that to the younger generation their absence rather than their presence is considered remarkable, has been almost wholly effected within the writer's memory, and is of itself sufficiently striking and important to justify the appellation of "The Wonderful Century" to that period which witnessed its rise, its progress, and its maturity of development.

# CHAPTER II

## LABOUR-SAVING MACHINERY

Wonderful chair! Wonderful houses! Wonderful people!
Whirr! whirr! all by wheels! Whizz! whizz! all by steam.
—*Eothen.*

Work—work—work
Till the brain begins to swim;
Work—work—work
Till the eyes are heavy and dim!
Seam and gusset and band,
Band and gusset and seam,
Till over the buttons I fall asleep,
And sew them on in a dream!
—*Hood.*

THE invention and partial development of much of our modern machinery dates from the last century, and our most advanced appliances for the manufacture of the various textile fabrics and hardware are mostly improvements of, or developments from, the older machines. These, taken in connection with the great improvements in steam-engines, have multiplied many times over the efficiency of human labour, but do not otherwise specially interest us here. There are, however, a few inventions which have the character of quite new departures, since not only do they greatly diminish labour, but they perform, by mechanical contrivances, operations which had been supposed to be beyond the power of machinery to execute.

The more prominent of these are the sewing machine, the typewriter, and the combined reaping, thrashing and winnowing machine, of which a brief account will be given.

The sewing machine, now so common, exercised the ingenuity of mechanicians for a long period before it arrived at sufficient perfection to be suitable for general use. The earlier machines were for embroidering only; then, about 1790, one was made for stitching shoes and other leather work, but it does not seem to have come into general use. A crocheting machine was patented in 1834; somewhat later one for rough basting; but it was not till 1846 that the first effective lock-stitch sewing machine was made by Elias Howe, of Cambridge, Massachusetts. Henceforth sewing machines were rapidly improved and adapted to every variety of work; but the difficulty of the problem to be solved is shown by the unusually long process of gradual development, much of the mechanical talent of both hemispheres being occupied for nearly a century before the various machines so familiar to-day were perfected. There are now special machines for making button-holes and for sewing on buttons, for carpet-sewing, for pattern-sewing, for leather work, and for the special operations required in the making and repairing of shoes. Boot and shoe-making by machinery, in large factories, has entirely grown up since the sewing-machine was proved to be adapted for almost every kind of sewing-work. As a result, machine-made boots and shoes are very cheap, but they are usually of inferior quality to the old hand-made articles; and first-class work is quite as dear as it was fifty or sixty years ago, or even dearer.

The typewriter is a still later invention, and though perhaps less difficult than the sewing-machine, yet it involves more complex motions and adjustments, so that the perfection it has so quickly attained is very remarkable. If we consider that about sixty separate types, including small letters, capitals, spaces, stops, etc., have to be so arranged and so connected as to be brought in any order whatever to a definite position, so as to form the successive letters and spaces in lines of printed characters, and then, being properly inked, must be brought into contact with the paper so as to produce a clear impression, and that all the motions of the machinery required must be the result of a single pressure on a key for each letter, following one another as rapidly as possible, we shall have some idea of the difficulties which have had to be overcome. Yet, so great are the resources of modern mechanism, and the ingenuity of our mechanists, that the required result has been attained in many different ways, so that we may now choose between half a dozen forms of typewriters, no one of which seems to be very markedly superior to the rest.

More important, perhaps, to mankind generally, are the harvesting machines, which render it possible to utilise one or two fine days to secure a harvest. Reaping machines have long been used in this country, and they were followed by combined reapers and binders, which left the crop ready for carting to the barn. But this, when the distance was great, did not save the grain from injury by wet, besides requiring much labour and a careful process of stacking to preserve it. In America a harvesting machine has been brought to perfection, which not only reaps the grain, but threshes it, winnows it, and delivers it into sacks ready for the granary or the market, at one operation. This machine, with two men, will, in one fine day, secure the crop from ten or fifteen acres, with a minimum of labour. In the great wheat-fields of California and Australia, with an almost uniformly dry climate at harvest time, it is this saving of labour which is the chief consideration; but in our treacherous climate, where a few days' delay may mean the partial or complete ruin of the crop, such machines will be doubly valuable by enabling farmers to utilize to the utmost every fine day after the grain is ripe. I had the pleasure of seeing this wonderful machine at work in California in 1887. It was propelled by sixteen small mules harnessed behind, so as not to be in the way; but steam power is now used. Considering what it effected, it was wonderfully light, compact and simple; and when agriculture is treated as a work of national importance, such machines will render us, to a considerable extent, independent of the weather, and will therefore become a necessity.

The three mechanical inventions here briefly described were conceived in the first half, and brought to perfection in the second half of the century. They each mark a new departure in human industry, inasmuch as they effect, by means of machinery, and at one operation, what had previously been performed by human labour directed by a hand or arm rendered skilful by long practice, and sometimes requiring several distinct operations. They had been thus performed during the whole preceding period of human history, or so long as the particular kind of work had been done; so that though of less general use and of less importance, they have the same distinguishing features which we have found to characterize our new methods of locomotion.

There are, of course, innumerable other remarkable mechanical inventions of the century in almost every department of industry—such as the Jacquard loom for pattern-weaving, revolvers and machine-guns, iron ships, screw-propellers, etc.; while machinery has been extensively applied to watch-making, screw-cutting, nail-making, printing, and a hundred other purposes. But none of these are of very high importance in themselves, or possess the special characteristics of being new and quite distinct departures from what has been done before, and they cannot therefore rank individually among those greater discoveries which pre-eminently distinguish the Nineteenth Century.

# CHAPTER III

## THE CONVEYANCE OF THOUGHT

Speak the word and think the thought,
Quick 'tis as with lightning caught—
Over, under, lands or seas
To the far antipodes.
* * * * *
I sent a message to my dear—
A thousand leagues and more to Her—
The dumb sea-levels thrilled to hear,
And Lost Atlantis bore to Her.

—*Kipling.*

THE history of the progress of communication between persons at a distance from each other, has gone through three stages which are radically distinct. At first it was dependent on the voice or on gestures, and a message to a friend (or enemy) at a distance could only be sent through a messenger, and was liable to distortion through failure of memory. The heralds and ambassadors of early times thus communicated orders from kings to their subjects, or conveyed messages from one king to another. Then came the invention of writing, and a new era of communication began. Letters were capable of conveying secret information and copious details, which could not be safely entrusted to the uncertain memory of an intermediary; and a single messenger could convey a large number of letters to various persons on the way to his ultimate destination. Henceforth the progress of communications was bound up with that of locomotion, and, as civilization advanced, arrangements were made for the conveyance of letters at a comparatively small cost. A post office for the public service was first established by some Continental merchants in the fourteenth century; but it was not till the time of Charles I. that anything of the kind was to be found in England, and then it was mainly for the purpose of keeping up a communication between London and Edinburgh, and the intervening large towns, for Government purposes. It was, however, the starting-point of our existing postal system, which has been gradually extended under the direction of the King's Postmaster-General, and has continued to be a Government monopoly to our day. The letters were carried on horseback till 1783, when mail coaches were first introduced; and these led to a great improvement in our main roads, and the extension of the postal service to every town and village in the kingdom.

But even with good roads and mail coaches, the actual time taken in the despatch of a letter to a distant place was little if any less than had been possible from the earliest times, by means of relays of runners on foot or by swift horsemen. The improvement consisted in the regularity and economy of the postal service. The introduction of railways and steamships enabled much greater speed to be secured; but the greatest and most beneficial improvement in the administration of the Post Office was that inaugurated by Rowland Hill in 1840. The rule, then first introduced, of an uniform charge irrespective of distance, is one of those entirely new departures so many of which characterize our century, and which not only produce immediate beneficial effects, but are the starting-points of various unforeseen developments. It was founded in this case on a careful estimate of the various items which make up the cost of the carriage and delivery of each letter, and it was shown that the actual conveyance, even for the greatest distances, was the smallest part of the cost when the number of letters is large, the chief items of expense being the office work—the sorting, stamping, packing, etc.—and the final delivery, all of which are quite independent of the distance the letter is carried. The old system, therefore, of increasing the charge for postage in proportion to distance was altogether unreasonable, because the cost of conveyance was hardly perceptibly increased; and if the Post Office was considered to be a public service for the public benefit only, the people had a right to demand that they should pay only in proportion to the cost. Yet the principle was not at first, and is not even now, fully carried out. For thirty years, from 1840 to 1871, the postage was increased equally with each successive increment of weight, the half-ounce letter being a penny, while one of two ounces was fourpence. But as the chief items of expense—the office work and delivery—were the same, or nearly the same, in both cases, the double or quadruple charge was entirely opposed to the principle on which the uniform rate was originally founded. Accordingly, in 1871, when an ounce letter was first carried for a penny, the charge for two ounces was fixed at three halfpence, while four ounces was taken for twopence. This accepted and common-sense principle, however, has not yet been applied to the charges of the Postal Union, so that a letter which is a fraction over the half-ounce is charged fivepence, or double, and one over an ounce and a half tenpence, or four times that of the half-ounce letter, although an extra halfpenny would probably cover the extra cost of the service in both cases.

The same inability of the official mind to carry out an admitted principle is seen also in the case of Postal Orders. The cost to the Post Office of receiving and paying money is exactly the same whether the amount is eighteen-pence or fifteen shillings, and there is neither justice nor common-sense in charging three times as much in the latter case. There is no risk, because the money is paid in advance; and as the amounts taken in and paid out for Postal Orders must be approximately equal, it is difficult to see what justification there is for making any difference in charge. The same objection applies to Money Orders; and as there is doubtless a certain percentage of both which, from various causes, are never presented for payment, the profit to the Post Office must be greater in case of the higher amounts, which is another reason why these should not be exceptionally taxed. When the railways are taken over by the State, to be worked for the good of the community only, the principle will admit of great extension, each increment of distance being charged at a lower rate, just as is each increment of weight in our inland letters.

The third stage in the means of communication, when by means of electric signals it was rendered independent of locomotion, is that which has especially distinguished the present century. The electric telegraph serves us as a new sense, enabling us to communicate with friends at the other side of the globe almost as rapidly and as easily as if they were in different parts of the same town. The means of communication we now use daily would have been wholly inconceivable to our ancestors a hundred years ago.

About the middle of the last century it was perceived by a few students of electricity that it afforded a means of communication at a distance; but it was not till the year 1837 that the efforts of many simultaneous workers overcame the numerous practical difficulties, and the first electric telegraph was established. Its utility was so great, especially in the working of the railways then being rapidly extended over the kingdom, that it soon came into general use; but hardly any one at first thought that it would ever be possible to lay wires across the ocean depths to distant continents. Yet, step by step, with wonderful rapidity, even this was accomplished. The first submarine line was laid from Dover to Calais in 1851; and only five years afterwards, in 1856, a company was formed to lay an electric cable across the Atlantic. The cable, 2,500 miles long and weighing a ton per mile, was successfully laid, in 1858, from Ireland to Newfoundland; but owing to the weakness of the electric current, and perhaps to imperfections in the cable, it soon became useless, and had to be abandoned. After eight years more of invention and experiment, another cable was successfully laid in 1866; and there are now no less than fourteen lines across the Atlantic, while all the other oceans have been electrically bridged, so that messages can be sent to almost any part of the globe at a speed which far surpasses the imaginary power of Shakespeare's sprite Ariel, who boasted that he could "put a girdle round about the earth in forty minutes." We are now able to receive accounts of great events almost while they are happening on the other side of the globe; and, owing to difference of longitude, we sometimes can hear of an event apparently before it has happened. If some great official were to die at Calcutta at sunset, we should receive the news soon after noon on the same day.

As a result of the numerous experimental researches necessitated for the continuous improvement of the electric telegraph, the telephone was invented, an even more marvellous and unexpected discovery. By it, the human voice, in all its countless modifications of quality and musical tone, and its most complex modulations during speech, is so reproduced at a distance that a speaker or singer can be distinctly heard and understood hundreds of miles away. This is not an actual transmission of the voice, as in the case of a speaking-tube, but a true reproduction by means of two vibrating discs, the one set in motion by the speaker, while the electric current causes identical vibrations in the similar disc at the end of the line, and these vibrations reproduce the exact tones of the voice so as to be perfectly intelligible. At first telephones could only be worked successfully for short distances, but by continuous improvements the distance has been steadily increased, so that in America there is a telephone line now in operation between New York and Chicago, cities about a thousand miles apart.

Those who have read Mr. Bellamy's wonderful story, *Looking Backward*, will remember the concerts continually going on day and night, with telephonic connections to every house, so that every one could listen to the very best obtainable music at will. But few persons are aware that a somewhat similar use of the telephone is actually in operation at Buda Pesth in the form of a telephonic newspaper. At certain fixed hours throughout the day a good reader is employed to send definite classes of news along the wires which are laid to subscribers' houses and offices, so that each person is able to hear the particular items he desires, without the delay of its being printed and circulated in successive editions of a newspaper. It is stated that the news is supplied to subscribers in this way at little more than the cost of a daily newspaper, and that it is a complete success.

We thus see that during the present century two distinct modes of communication with persons at a distance have been discovered and brought into practical use, both of which are perfectly new departures from the methods which, with but slight modifications, had been in use since that early period when picture-writing or hieroglyphics were first invented.

In the facilities and possibilities of communication with our fellow-men all over the world, the advance made in the present century is not only immensely greater than that effected during the whole preceding period of human history, but is even more marvellous in its results. And it is also much greater in amount, than the almost simultaneous advance in facilities for locomotion, great as these have been.

# CHAPTER IV

## FIRE AND LIGHT

> Put out the light, and then—Put out the light!
> If I quench thee, thou flaming minister,
> I can again thy former light restore,
> Should I repent me:—but once put out thy light,
> Thou cunning'st pattern of excelling nature,
> I know not where is that Promethean heat
> That can thy light relume.
>
> —*Othello.*

It seems probable that the discovery of the use of fire, and of some mode of producing it at will, constituted the first important advance of primitive man towards obtaining that command over nature which we term civilisation. How long ago it is since that first step was taken, we have no means of determining. The palæolithic cave-dwellers made use of fire, and no tribes of men have been found who were wholly unacquainted with it. It was probably first utilised in volcanic districts where sticks may often be ignited by thrusting them into cavities in old lava streams. In other regions, trees are often ignited when struck by lightning; and when this was first observed, the agreeable warmth, the ease with which the fire could be kept up by adding fresh fuel, the cheerful blaze at night, and the pleasant taste imparted to many kinds of food by roasting, would almost certainly lead to its careful preservation, and its distribution to other families and tribes. When once used, the inconvenience of losing it would be so great, that any clue to its mode of production would be followed up. It is said that trees are sometimes ignited by the friction of dry branches which happen to touch each other, when violently rubbed together during a strong wind. When this was observed for the first time by some thoughtful savage, and he actually found that strong rubbing did make things hot, he would be encouraged to use his utmost efforts to imitate the effect produced by nature. After many unsuccessful trials, he would at length succeed; and the important news would be rapidly communicated to adjacent tribes, and thus spread over a whole continent. As a matter of fact, this method of producing fire by friction is that most common among savages in all parts of the world; and since it requires only materials that are almost everywhere at hand, it descended even to some civilised peoples. It is, however, a rather troublesome process, requiring a considerable amount of skill and perseverance; hence some of the lowest savages, such as the Tasmanians, are said to have been without the knowledge of it, keeping their fires constantly alight, or, when accidently extinguished, obtaining it from some adjacent tribe. Perhaps, however, the dampness of their forests rendered it practicable only during very dry seasons.

The more convenient method of striking a light by the use of flint, steel, and tinder, probably originated after iron was first made, and soon became adopted by all civilised people, and by many savages who possessed iron; and this method continued in use from the times of prehistoric man through all the ages of barbarism and civilisation down to the early part of this century, and the process underwent hardly any improvement during that long period. One of the most vivid recollections of my childhood is of seeing the cook make tinder in the evening, by burning old linen rags, and in the morning, with flint and steel, obtain the spark which, by careful blowing, spread sufficiently to ignite the thin brimstone match from which a candle was lit and fire secured for the day. The process was, however, sometimes, a tedious one; and if the tinder had accidently got damp, or if the flint were worn out, after repeated failures a light had to be obtained from a neighbour. At that time there were few savages in any part of the world but could obtain fire as easily as the most civilised of mankind.

At length, after the use of these rude methods for many thousand years, a great discovery was made which revolutionised the process of fire-getting. The properties of phosphorus were known to the alchemists, and it is strange that its ready ignition by friction was not made use of to obtain fire at a much earlier period. It was, however, both an expensive and a dangerous material, and though about a hundred years ago it began to be made cheaply from bones, it was not used in the earliest friction matches. These were invented in 1827, or a little earlier, by Mr. John Walker, a chemist and druggist of Stockton-on-Tees, and consisted of wood splints dipped in chlorate of potash and sulphur mixed with gum, which ignited when rubbed on sandpaper. Two years later the late Sir Isaac Holden invented a similar match. About 1834, phosphorus began to be used with the other materials to cause more easy ignition, and by 1840 these matches became so cheap as to come into general use in place of the old flint and steel. They have since spread to every part of the world, and their production constitutes one of the large manufacturing industries of England, Sweden, and many other countries.

Here again we have an invention that is not a modification of the older mode of obtaining fire, but a new departure, possessing such great advantages that it rapidly led to the almost total abandonment of the old methods in every civilised country, as well as in many of the remotest and least civilised parts of the world. For many thousands of years the means of obtaining fire remained almost unchanged over the whole world, till only sixty years ago, a discovery which at the time seemed of but little importance, has led to a quicker and easier process, which is so widely adopted that millions of persons in all civilised countries make use of it every day of their lives.

Coming now to the use of fire as a light-giver, we find that an even greater change has taken place in our time. The first illuminants were probably torches made of resinous woods which will give a flame for a considerable time. Then the resin exuding from many kinds of trees would be collected and applied to sticks or twigs, or to some fibrous materials tied up in bundles, such as are still used by many savage peoples, and were used in the old baronial halls. For out-door lights torches were used almost down to our times, an indication of which is seen in the iron torch-extinguishers at the doors of many of the older West End houses; while, before the introduction of gas, link-boys were as common in the streets as match-sellers are now. Then came lamps, formed of small clay cups, holding some melted animal fat and a fibrous wick; and, somewhat later, rushlights and candles. Still later, vegetable oils were used for lamps, and wax for candles; but the three modes of obtaining illumination for domestic purposes remained entirely unchanged in principle, and very little improved, throughout the whole period of history down to the end of the eighteenth century. The Greek and Roman lamps, though in beautiful receptacles of bronze or silver, were exactly the same in principle as those of the lowest savage, and hardly better in light-giving power; and though various improvements in form were introduced, the first really important advance was made by the Argand burner. This introduced a current of air into the centre of the flame as well as outside it, and, by means of a glass chimney, a regular supply of air was kept up, and a steady light produced. Although the invention was made at the end of the last century the lamps were not sufficiently improved and cheapened to come into use till about 1830; and from that time onward many other improvements were made, chiefly dependent on the use of the cheap mineral oils, rendering lamps so inexpensive, and producing so good a light, that they are now found in the poorest cottages.

The only important improvement in candles is due to the use of paraffin fats instead of tallow, and of flat plaited wicks which are consumed by the flame. In my boyhood, the now extinct "snuffers" were in universal use, from the common rough iron article in the kitchen to elaborate polished steel spring-snuffers of various makes for the parlour, with pretty metal or *papier-maché* trays for them to stand in. Candles are still very largely used, being more portable and safer than most of the paraffin oil lamps. Even our lighthouses used only candles down to the early part of the present century.

A far more important and more radical change in our modes of illumination was the introduction of gas-lighting. A few houses and factories were lighted with gas at the very end of the last century, but its first application to out-door or general purposes was in 1813, when Westminster Bridge was illuminated by it, and so successfully that its use rapidly spread to every town in the kingdom, for lighting private houses as well as streets and public buildings. When it was first proposed to light London with gas, Sir Humphry Davy is said to have declared it to be impracticable, both on account of the enormous size of the needful gasholders, and the great danger of explosions. These difficulties, have, however, been overcome, as was the supposed insuperable difficulty of carrying sufficient coal in the case of steamships crossing the Atlantic, the impossibilities to one generation becoming the realities of the next.

Still more recent, and more completely new in principle, is the electric light, which has already attained a considerable extension for public and private illumination, while it is applicable to many purposes unattainable by other kinds of light. Small incandescent lamps are now used for examinations of the larynx and in dentistry, and a lamp has even been introduced into the stomach by which the condition of that organ can be examined. For this last purpose numerous ingenious arrangements have to be made to prevent possible injury, and by means of prisms at the bends of the tube the operator can inspect the interior of the organ under a brilliant light. Other internal organs have been explored in a similar manner, and many new applications in this direction will no doubt be made. In illuminating submarine boats and exploring the interiors of sunken vessels it does what could hardly be effected by any other means.

We thus find, that whereas down to the end of the last century our modes of producing and utilising light were almost exactly the same as had been in use for the preceding two or three thousand years, in the present century we have made no less than three new departures, all of which are far superior to the methods of our forefathers. These are—(1) the improvement in lamps by the use of the principle of the Argand burner and chimney; (2) lighting by coal-gas; and (3) the various modes of electric lighting. The amount of advance in this one department of domestic and public illumination during the present century is enormous, while the electric light has opened up new fields of scientific exploration.

Whether we consider the novelty of the principles involved, or the ingenuity displayed in their application, we cannot estimate this advance at less than that effected during the whole preceding period of human history, from that very remote epoch when fire was first taken into the service of mankind, down to the time of men now living among us.

# CHAPTER V

## NEW APPLICATIONS OF LIGHT—PHOTOGRAPHY

O portrait, bright and wonderful!
Wrought by the sun-god's pencil true;
What grace of feature, glance of eye!
The soul itself beams out from you.

New marvel of a marvellous age!
Apelles old, whose art 'twas said
Rivalled reality, than this
Had never limned a lovelier head.[1]

THE improvements in the mode of production of light for common use, discussed in the previous chapter, are sufficiently new and remarkable to distinguish this century from all the ages that preceded it, but they sink into insignificance when compared with the discoveries which have been made as to the nature of light itself, its effects on various kinds of matter leading to the art of Photography, and the complex nature of the Solar Spectrum leading to Spectrum Analysis. This group of investigations alone are sufficient to distinguish the present century as an epoch of the most marvellous scientific discovery.

[1] The above translation of the Pope's Latin verse on Photography, is by my friend, Mr. F. T. Mott, of Leicester.

Expressa solis spiculo
Nitens imago, quambene
Frontis decus, vim luminum
Refers, et oris gratiam.

O mira virtus ingeni,
Novumque monstrum! Imaginem
Naturæ Apelles æmulus
Non pulchriorem pingeret.

Although Huygens put forward the wave-theory of light more than two hundred years ago, it was not accepted, or seriously studied, till the beginning of the present century, when it was revived by Thomas Young, and was shown by himself, by Fresnel, and other mathematicians, to explain all the phenomena of refraction, double-refraction, polarisation, diffraction, and interference, some of which were inexplicable on the Newtonian theory of the emission of material particles, which had previously been almost universally accepted. The complete establishment of the undulatory theory of light is a fact of the highest importance, and will take a very high place among the purely scientific discoveries of the century.

From a more practical point of view, however, nothing can surpass in interest and importance the discovery and continuous improvement of the photographic art, which has now reached such a development, that there is hardly any science or any branch of intellectual study that is not indebted to it. A brief sketch of its origin and progress will therefore not be uninteresting.

The fact that certain salts of silver were darkened by exposure to sunlight, was known to the alchemists in the sixteenth century, and this observation forms the rudiment from which the whole art has been developed. The application of this fact to the production of pictures belongs, however, wholly to our own time. In the year 1802, Wedgewood described a mode of copying paintings on glass by exposure to light, but neither he nor Sir Humphrey Davy could find any means of rendering the copies permanent. This was first effected in 1814 by M. Niepce of Châlon, but no important results were obtained till 1839, when

Daguerre perfected the beautiful process known as the daguerrotype. Permanent portraits were taken by him on silvered plates, and they were so delicate and beautiful that probably nothing in modern photography can surpass them. For several years they were the only portraits taken by the agency of light, but they were very costly, and were therefore completely superseded when cheaper methods were discovered.

About the same time a method was found for photographing leaves, lace, and other semi-transparent objects on paper, and rendering them permanent, but this was of comparatively little value. In the year 1850, the far superior collodion-film on glass was perfected, and negatives were taken in a camera-obscura, which, when placed on black velvet, or when coated with a black composition, produced pictures almost as perfect and beautiful as the daguerrotype itself, and at much less cost. Soon afterwards positives were printed from the transparent negatives on suitably prepared paper, and thus was initiated the process, which, with endless modifications and improvements, is still in use. The main advance has been in the increased sensitiveness of the photographic plates, so that, first, moving crowds, then breaking waves, running horses, and other quickly moving objects were taken, while now a bullet fired from a rifle can be photographed in the air.

With such marvellous powers, photography has come to the aid of the arts and sciences in ways which would have been perfectly inconceivable to our most learned men of a century ago. It furnishes the Meteorologist, the Physicist, and the Biologist, with self-registering instruments of extreme delicacy, and enables them to perserve accurate records of the most fleeting natural phenomena. By means of successive photographs at short intervals of time, we are able to study the motions of the wings of birds, and thus learn something of the mechanism of flight; while even the instantaneous lightning-flash can be depicted, and we thus learn, for the first time, the exact nature of its path.

Perhaps the most marvellous of all its achievements is in the field of astronomy. Every increase in the size and power of the telescope has revealed to us ever more and more stars in every part of the heavens; but, by the aid of photography, stars are shown which no telescope that has been, or that probably ever will be constructed, can render visible to the human eye. For by exposing the photographic plate in the focus of the object glass for some hours, almost infinitely faint stars impress their image, and the modern photographic star-maps show us a surface densely packed with white points that seem almost as countless as the sands of the seashore. Yet every one of these points represents a star in its true relative position to the visible stars nearest to it, and thus gives at one operation an amount of accurate detail which could hardly be equalled by the labour of an astronomer for months or years—even if he could render all these stars visible, which, as we have seen, he cannot do. A photographic survey of the heavens is now in progress on one uniform system, which, when completed, will form a standard for future astronomers, and thus give to our successors some definite knowledge of the structure, and perhaps, of the extent of the stellar universe.

Within the last few years the mechanical processes

by means of which photographs can now be reproduced through the printing press have been rendered so perfect that books and periodicals are illustrated with an amount of accuracy and beauty that would have been impossible, even twenty years ago, except at a prohibitive cost.

It has long been the dream of photographers to discover some mode of obtaining pictures which shall reproduce all the colours of nature without the intervention of the artist's manipulation. This was seen to be exceedingly difficult, if not impossible, because the chemical action of coloured light has no power to produce pigments of the same colour as the light itself, without which a photograph in natural colours would seem to be impossible. Nevertheless, the problem has been solved, but in a totally different manner; that is, by the principle of "interference," instead of by that of chemical action. This principle was discovered by Newton, and is exemplified in the colours of the soap bubble, and in those of mother-of-pearl and other iridescent objects. It depends on the fact that the differently coloured rays are of different wave-lengths, and the waves reflected from two surfaces half a wave-length apart neutralize each other and leave the remainder of the light coloured. If, therefore, each differently coloured ray of light can be made to produce a corresponding minute wave-structure in a photographic film, then each part of the film will reflect only light of that particular wave-length, and therefore of that particular colour, that produced it. This has actually been done by Professor Lippmann, of Paris, who published his method in 1891; and in a lecture before the Royal Society in April,

1896, he fully described it and exhibited many beautiful specimens.[1]

The method is as follows:—A sensitive film, of some of the usual salts of silver in albumen or gelatine, is used, but with much less silver than usual, so as to leave the film quite transparent. It must also be perfectly homogeneous, since any granular structure would interfere with the result. This film on glass must be placed in a frame so constructed that at the back of it there is a shallow cell that can be filled with mercury which is in contact with the film. It is then exposed in the usual way, but much longer than for an ordinary photograph, so that the light-waves have time to produce the required effect. The light of each particular tint being reflected by the mercury, meets the incoming light and produces a set of *standing* waves—that is, of waves surging up and down, each in a fixed plane. The result is, that the metallic particles in the film become assorted and stratified by this continued wave-action, the distance apart of the strata being determined by the wave-length of the particular coloured light—for the violet rays about eight millionths of an inch; so that in a film of ordinary thickness there would be about five hundred of these strata of thinly scattered metallic particles. The quantity of silver used being very small, when the film is developed and fixed in the usual way the result is not a light-and-shade negative, but a nearly transparent film which nevertheless reflects a sufficient amount of light to produce a naturally coloured picture.

The principle is the same for the light-waves as that of the telephone for sound-waves. The voice sets

[1] This lecture is reported in *Nature*, vol. liii. p. 617.

up vibrations in the transmitting diaphragm, which, by means of an electric current, are so exactly reproduced in the receiving diaphragm as to give out the same succession of sounds. An even more striking and, perhaps, closer analogy is that of the phonograph, where the vibrations of the diaphragm are permanently registered on a wax cylinder, which, at any future time, can be made to set up the same vibrations of the air, and thus reproduce the same succession of sounds, whether words or musical notes. So, the rays of every colour and tint that fall upon the plate throw the deposited silver within the film into minute strata which permanently reflect light of the very same wave-length, and therefore of the very same colour as that which produced them.

The effects are said to be most beautiful, the only fault being that the colours are more brilliant than in nature, just as they are when viewed in the camera itself. This, however, may perhaps be remedied (if it requires remedying) by the use of a slightly opaque varnish. The comparatively little attention that has been given to this beautiful and scientifically-perfect process, is no doubt due to the fact that it is rather expensive, and that the pictures cannot, at present, be multiplied rapidly. But for that very reason it ought to be especially attractive to amateurs, who would have the pleasure of obtaining exquisite pictures which will not become commonplace by indefinite reproduction.

The brief sketch of the rise and progress of photography now given, illustrates the same fact which we have already dwelt upon in the case of other discoveries. This beautiful and wonderful art, which already plays an important part in the daily life and enjoyment of all civilized people, and which has extended the bounds of human knowledge into the remotest depths of the starry universe, is not an improvement of, or development from, anything that went before it, but is a totally new departure. From that early period when the men of the stone age rudely outlined the mammoth and the reindeer on stone or ivory, the only means of representing men and animals, natural scenery, or the great events of human history, had been through the art of the painter or the sculptor. It is true that the highest Greek, or Mediæval, or Modern art, cannot be equalled by the productions of the photographic camera; but great artists are few and far between, and the ordinary, or even the talented draughtsman can give us only suggestions of what he sees, so modified by his peculiar mannerism as often to result in a mere caricature of the truth. Should some historian in Japan study the characteristics of English ladies at two not remote epochs, as represented, say, by Frith and by Du Maurier, he would be driven to the conclusion that there had been a complete change of type, due to the introduction of some foreign race, in the interval between the works of these two artists. From such errors as this we shall be saved by photography; and our descendants in the middle of the coming century will be able to see how much, and what kind, of change really does occur from age to age.

The importance of this is well seen by comparing any of the early works on Ethnology, illustrated by portraits intended to represent the different "types of mankind," with recent volumes which give us copies of actual photographs of the same types; when we shall see how untrue to nature are the former, due probably to the artist having delineated those extreme forms, either of ugliness or of beauty, that most attracted his attention, and to his having exaggerated even these. Thus only can we account for the pictures in some old voyages, showing an English sailor and a Patagonian as a dwarf beside a giant; and for the statement by the historian of Magellan's voyage, that their tallest sailor only came up to the waist of the first man they met. It is now known that the average height of Patagonian men is about five feet ten inches, or five feet eleven inches, and none have been found to exceed six feet four inches. Photography would have saved us from such an error as this.

There will always be work for good artists, especially in the domain of colour and of historical design; but the humblest photographer is now able to preserve for us, and for future generations, minutely accurate records of scenes in distant lands, of the ruins of ancient temples which are sometimes the only record of vanished races, and of animals or plants that are rapidly disappearing through the agency of man. And, what is still more important, they can preserve for us the forms and faces of the many lower races which are slowly but surely dying out before the rude incursions of our imperfect civilization.

That such a new and important art as photography should have had its birth, and have come to maturity, so closely coincident with the other great discoveries of the century already alluded to, is surely a very marvellous fact, and one which will seem more extraordinary to the future historian than it does to ourselves, who have witnessed the whole process of its growth and development.

The most recent of all the discoveries in connection with light and photography, and one which extends our powers of vision in a direction and to an extent the limits of which cannot yet be guessed at, is that peculiar form of radiation termed the X, or Röntgen Rays, from Professor Röntgen, of Würzburg, who was the first to investigate their properties and make practical applications of them. These rays are produced by a special form of electrical current sent through a vacuum tube, in or around which is some fluorescent substance, which under the action of the current become intensely luminous. But this luminosity has totally different properties from ordinary light, inasmuch as the substances which are opaque or transparent to it are not the same as those to which we usually apply the terms, but often the very contrary. Paper, for instance, is so transparent that the rays will pass through a book of a thousand pages, or through two packs of cards, both of which would be absolutely opaque to the most brilliant ordinary light. Aluminium, tin, and glass of the same thickness are all transparent, but they keep out a portion of the rays; whereas platinum and lead are quite opaque. To these rays aluminium is two hundred times as transparent as platinum. Wood, carbon, leather and slate are much more transparent to the X Rays than is glass, some kinds of glass being almost opaque, though quite transparent to ordinary light. Flesh and skin are transparent in moderate thicknesses, while bone is opaque. Hence, if the rays are passed through the hand the bones cast a shadow, though an invisible one; and as, most fortunately, the rays act upon photographic plates almost like ordinary light, hands or other parts of the body can be photographed by their shadows, and will show the bones by a much darker tint. Hence their use in surgery, to detect the exact position of bullets or other objects embedded in the flesh or bone. A needle which penetrated the knee joint and then broke off, leaving a portion embedded which set up inflammation, and might have necessitated the loss of the limb, was shown so accurately that a surgeon cut down to it and got it out without difficulty.

An exceptional property of these rays is, that they cannot be either refracted or reflected as can ordinary light and heat. Hence it is only the shadow that can be photographed. And another curious result of this is that they can pass through a powder as easily as through a solid; whereas ordinary light cannot pass through powdered glass or ice, owing to the innumerable reflections and refractions which soon absorb all the rays except those reflected from a very thin surface layer. Proportionate thicknesses of aluminium or zinc, whether in the solid plate or in powder are equally transparent to these singular rays.

So much is already popularly known on this subject that it is not necessary to go into further details here. But this new form of radiant energy opens up so many possibilities, both as to its own nature and as to the illimitable field of research into the properties and powers of the mysterious ether, that it forms a fitting and dramatic climax to the scientific discoveries of the century.

# CHAPTER VI

## NEW APPLICATIONS OF LIGHT:—SPECTRUM ANALYSIS

Far beyond Orion bright
  Cloud on cloud the star-haze lies;
Million years bear down the light
  Earthward from those ghost-like eyes.

—*F. T. Palgrave.*

Hushed be all earthly rhymes!
  List to those spheral chimes
That echo down the singing vaults of night.
  The quivering impulse runs
  From the exultant suns,
Circling in endless harmonies of light.

—*F. T. Mott.*

Among the numerous scientific discoveries of our century, we must give a very high, perhaps even the highest, place to Spectrum Analysis. Not only because it has completely solved the problem of the true nature and cause of the various spectra produced by different kinds of light, but because it has given us a perfectly new engine of research, by which we are enabled to penetrate into the remotest depths of space, and learn something of the constitution and the motions of the constituent bodies of the stellar universe. Through its means we have acquired what are really the equivalents of new senses, which give us knowledge that before seemed absolutely and for ever unattainable by man.

The solar spectrum is that coloured band produced by allowing a sunbeam to pass through a prism, and a portion of it is given by the dewdrop or the crystal when the sun shines upon them; while the complete band is produced by the numerous raindrops, the coloured rays from which form the rainbow. Newton examined the colours of the spectrum very carefully, and explained them on the theory that light of different colours has different refrangibilities—or, as we now say, different wave-lengths. He also showed that a similar set of colours can be produced by the interference of rays when reflected from the two surfaces of very thin plates, as in the case of what are termed Newton's rings and in the iridescent colours of thin films of oil on water, of soap bubbles, and many other substances.

These colour-phenomena, although very interesting in themselves, and giving us more correct ideas of the nature of colour in the objects around us, did not lead to anything further. But in 1802, the celebrated chemist, Dr. Wollaston, made the remarkable discovery that the solar-spectrum, when closely examined, is crossed by very numerous black lines of various thicknesses, and at irregular distances from each other. Later, in 1817, these lines were carefully measured and mapped by Fraunhofer; but their meaning remained an unsolved problem till about the year 1860, when the German physicist, Kirchhoff, discovered the secret, and opened up to chemists and astronomers a new engine of research whose powers are probably not yet exhausted.

It was already known that the various chemical elements, when heated to incandescence, produce spectra consisting of a group of coloured bands, and it had been noticed that some of these bands, as the yellow band of Sodium, corresponded in position with certain black lines in the solar spectrum. Kirchhoff's discovery consisted in showing that, when the light from an incandescent body passes through the same substance in a state of vapour, much of it is absorbed, and the coloured bands become replaced by black lines. The black lines in the solar spectrum are due, on this theory, to the light from the incandescent body of the sun being partially absorbed in passing through the vapours which surround it. This theory led to a careful examination of the spectra of all the known elements, and on comparing them with the solar spectrum it was found that in many cases the coloured bands of the elements corresponded exactly in position with certain groups of black lines in the solar spectrum. Thus hydrogen, sodium, iron, magnesium, copper, zinc, calcium, and many other elements have been proved to exist in the sun. Some outstanding solar lines, which did not correspond to any known terrestrial element, were supposed to indicate an element peculiar to the sun, which was therefore named Helium. Quite recently this element has been discovered in a rare mineral, and its coloured spectrum is found to correspond exactly to the dark lines in the solar spectrum on which it was founded, thus adding a final proof of the correctness of the theory, and affording a striking example of its value as an instrument of research.

The immediate effect of the application of the spectroscope to the stars was very striking. The supposition that they were suns became a certainty, since they gave spectra similar in character and often very closely resembling in detail that of our sun. Aldebaran is one of the most sun-like stars, being yellow in colour and possessing lines which indicate most of the elements found in the sun. White stars, such as Sirius and Vega, show hydrogen lines only; and these are supposed to be hotter than our sun, and in an earlier stage of development, while red stars are supposed to be cooling. Other explanations of these facts have, however, been suggested. Much information has also been obtained as to the nature of the nebulæ. Sir William Herschel supposed that they were all really star-clusters, but so enormously remote that even the most powerful telescopes could not render visible the stars composing them. Later observations have shown that many of them do consist of stars, or star-dust, as it has been called; and this seemed to support the theory that all were so composed, including the milky way. A study of the distribution of stars and nebulæ by Proctor and others led, however, to the conclusion that they were often really connected, and that nebulæ were not, on the average, more distant than stars; and this view has been confirmed by the spectroscope, which has shown them often to consist of glowing gas; and this is especially the characteristic of all those situated in or near the milky way. The first great result of spectrum-analysis has thus been to demonstrate the real nature of many stars and nebulæ, to determine some of the elements of which they are formed, and to give us some indications of the changes they have undergone, and thus help us towards a general theory of the development of the stellar universe.

Marvellous as is this extension of our knowledge of objects, so distant that our largest telescopes are powerless to show them as more than points of light, it is only a part, perhaps only a small part, of what the spectroscope has already done, or may yet do, for astronomy. By a most refined series of observations it has enabled us to detect and measure certain motions of the stars which seemed to be wholly beyond our grasp, and also to demonstrate the existence of celestial bodies which could be detected in no other way.

In order to understand how this is possible we have to make use of the wave-theory of light; and the analogy of other wave-motions will enable us better to grasp the principle on which these calculations depend. If on a nearly calm day we count the waves that pass each minute by an anchored steam-boat, and then travel in the direction the waves come from, we shall find that a larger number pass us in the same time. Again, if we are standing near a railway, and an engine comes towards us whistling, we shall notice that it changes its tone as it passes us; and as it recedes the sound will be very different, although the engine is at the same distance from us as when it was approaching. Yet the sound does not change to the ear of the engine-driver, the cause of the change being that the sound-waves reach us in quicker succession as the source of the waves is approaching us than when it is retreating from us. Now just as the pitch of a note depends upon the rapidity with which the air-vibrations reach our ear, so does the colour of a particular part of the spectrum depend upon the rapidity with which the ethereal waves which produce colour reach our eyes; and as this rapidity is greater when the source of the light is approaching than when it is receding from us, a slight shifting of the position of the dark lines will occur, as compared with their position in the spectrum of the sun or of any stationary source of light, if there is any motion sufficient in amount to produce a perceptible shift. On experimenting with a powerful spectroscope constructed for the purpose, Sir William Huggins, in 1868, found that such a change did occur in the case of many stars, and that their rate of motion towards us or away from us—termed the radial motion—could be calculated. As the actual distance of some of these stars has been measured, and their change of position annually (their proper motion) determined, the additional factor of the amount of motion in the direction of our line of sight, completes the data required to fix their true line of motion among the other stars.

This method of research has now been applied to many double stars with great success, observations of their spectra showing that in some cases they move one towards and one away from us, as they must do if they are revolving around their common centre of gravity in an ellipse whose plane lies approximately in our direction. It has also brought to light the interesting fact that some stars which appear single in the most powerful telescopes are really double, since their spectra show a shifting of spectrum lines, which after a considerable time changes to an opposite direction, and by the period occupied in the complete change of direction the time of rotation of the component stars can be determined, although one of the components has never been seen. By this means the variable star Algol has been proved to have a dark companion which partially eclipses it every 69 hours; and both Sirius and Procyon have been shown to have dark or less visible companions, that of Sirius being really just visible in the very best telescopes. The unusual motions of Sirius have been long known, and were supposed to be due to the presence of a companion, which has now been shown to be the true explanation.

The accuracy of this method under favourable conditions is very great, as has been proved by those cases in which we have independent means of calculating the real motion. The motion of Venus towards or away from us can be calculated with great accuracy for any period, being a resultant of the combined motions of the planet and of our earth in their respective orbits. The radial motions of Venus were determined at the Lick Observatory in August and September, 1890, by spectroscopic observations, and also by calculation, to be as follows:—

| | BY OBSERVATION. | BY CALCULATION. |
|---|---|---|
| Aug. 16th. | — 7·3 miles per second. | — 8·1 miles per second. |
| „ 22nd. | — 8·9 „ | — 8·2 „ |
| „ 30th. | — 7·3 „ | — 8·3 „ |
| Sept. 3rd. | — 8·3 „ | — 8·3 „ |
| „ 4th. | — 8·2 „ | — 8·3 „ |

showing that the maximum error was only one mile per second, while the mean error was about a quarter of a mile. Owing to the greater difficulty in observing the spectra of stars, the accuracy in their case is probably not quite so great. This has been tested by observations of the same star at times when the earth's motion in its orbit is towards or away from the star, whose apparent radial velocity is, therefore, increased or diminished by a known amount. Observations of this kind were made by Dr. Vogel, Director of the Astrophysical Observatory at Potsdam, showing, in the case of three stars, of which ten observations were taken, a mean error of about two miles per second.

The same observer, from his study of the spectra of the variable star Algol, has been able to determine that both the visible star and its dark companion are somewhat larger than our sun, though of less density; that their centres are 3,230,000 miles apart, and that they move in their orbits at rates of 55 and 26 miles per second respectively; and this information, it must be remembered, has been gained, as to objects the light of which takes about forty-seven years to reach us!

So striking are these results, and so rapid has been the increase in the delicacy and trustworthiness of the observations, that the President of the Royal Astronomical Society, in an address delivered in 1893, contemplated the possibility that, by still further refinements in the application of the spectroscope, the most accurate measures of the rate of motion of our earth in its orbit, and, therefore, of the distance of the sun, might be deduced from observations of stars which are themselves so remote as to be beyond our powers of measurement.

So late as the year 1842 the French mathematician and philosopher, Comte, declared that all study of the fixed stars was waste of time, because their distance was so great that we could never learn anything about them—a striking illustration of the complete novelty, no less than of the wonderful possibilities of this marvellous engine of research. Not only is it a wholly new departure from anything known or even imagined before, but it is able to give us a large and varied amount of knowledge of that portion of the visible universe which has hitherto been the least known and which seemed to be the most hopelessly unapproachable. On every ground, therefore, we must place the discovery and applications of Spectrum Analysis as deserving of the highest place among the numerous great scientific achievements of the nineteenth century.[1]

[1] An admirable popular account of the application of the spectroscope to the heavens, will be found in an article on "The New Astronomy," in the *Nineteenth Century* of June, 1897. It is written by Sir William Huggins, the greater part of whose life has been devoted to this branch of the science, in which he was one of the earliest and most successful observers and discoverers.

# CHAPTER VII

## THEORETICAL DISCOVERIES IN PHYSICS

Has matter motion? Then each atom,
Asserting its perpetual right to dance,
Would make a universe of dust!

---

For the world was built in order,
And the atoms march in tune.

—*Emerson.*

THE theoretical discoveries in the domain of physics (besides those already referred to) have been very numerous, but only a few of them have enough generality or have become sufficiently popular to require notice in the present sketch. Two of these discoveries, however, stand above the general level as important contributions to our knowledge of the material universe. These are (1) the determination of the mechanical equivalent of heat leading to the general theory of the conservation of energy, and (2) the molecular theory of gases.

Down to the beginning of this century heat was generally considered to be a form of matter, termed caloric or phlogiston. The presence of phlogiston was supposed to render substances combustible, but when the chemical theory of combustion was discovered by Lavoisier, phlogiston, as the cause of combustion, disappeared, although caloric, as the material basis of heat, still held its ground. Close to the end of the last century Count Rumford showed, that in boring a brass cannon the heat developed in 2½ hours was sufficient to raise 26½ lbs. of water from the freezing to the boiling point. But, during the operation, the metal had lost no weight or undergone any other change; and as the production of heat by this process appeared to be unlimited, he concluded that heat could not be matter, but merely a kind of motion set up in the particles of matter by the force exerted. Bacon and Locke had expressed similar ideas long before; and, later, Sir Humphry Davy showed that by rubbing together two pieces of ice at a temperature below the freezing point sufficient heat was produced to partially melt them; while other observers found that to shake water in a bottle raised its temperature, and that percussion or compression, as had been long known, produced a considerable amount of heat. These various facts led to the conclusion that there was a mechanical equivalent of heat—that is, that a certain amount of force exerted or work done would produce a corresponding amount of heat; and Joule was the first to determine this accurately by a number of ingenious experiments. The result was found to be, that a pound of water can be raised 1° C. by an amount of work equal to that required to raise one pound to the height of 1,392 feet, or 1,392 lbs. one foot. Various experiments with different materials were found always to lead to the same result, and thus the final blow was given to the material theory of heat, which was thenceforth held to be *a mode of motion of the molecules of bodies.*

These conclusions led to the more general law of the conservation of energy, which implies that in any limited system of bodies, whether a steam-engine or the solar system, no change can occur in the total amount of the energy it contains unless fresh energy comes to it from without, or is lost by transmission to bodies outside it. But as, in the case of the sun, some heat is certainly lost by radiation into space, unless an equal amount comes in from the stellar universe, the system must be cooling, and in sufficient time would lose all its heat, and therefore much of its energy. The chief use of the principle is to teach us what becomes of force expended without any apparent result, as when a ball falls to the ground and comes to rest. We now know that the energy of the falling ball is converted into heat, which, if it could be all preserved and utilized, would again raise the ball to the height from which it fell. It also enables us to trace most of the energy around us, whether of wind, or water, or of living animals, to the heat and light of the sun. Wind is caused by inequalities of the sun's heat on the earth; all water power is due to evaporation by the sun's heat, which thus transfers the water from the ocean surface to the mountains, producing rivers; solar heat alone gives power to plants to absorb carbonic acid and build up their tissues, and the energy thus locked up is again liberated during the muscular action of the animals which have fed directly or indirectly on the plants.

This great principle enables us to realize the absolute interdependence of all the forces of nature. It teaches us that there is no origination of force upon the earth, but that all energy either now comes to us from the sun or was originated in the sun before our earth separated from it; and we are thus led to the conclusion, that all work, all motion, every manifestation of power we see around us, are alike the effects of heat or of other radiant forces allied to it. This conclusion we shall find is still further enforced by the next great discovery we have to notice.

*The Molecular Theory of Gases.*

The very remarkable properties of gases, their apparently unlimited elasticity and indefinite powers of expansion, were very difficult to explain on any theory of their molecules being subject to such attractive and repulsive forces as seem to exist in other states of matter. A consideration of these properties, together with the power of diffusion, by which gases of very different densities form a perfect mixture when in contact, and the fact that by the application of heat almost all liquids and many solids can be changed into gases, led to the conception that they owed their peculiar properties to their molecules being in a state of intensely rapid motion in all directions. On this theory the molecules are very far apart in proportion to their size, and are continually coming in contact with each other. Owing to their perfect elasticity, they rebound without loss of motion or energy, and their continual impact against the sides of the vessel containing them is what gives to gases their great expansibility. From a study of these various properties it has been calculated that, at ordinary temperatures, there are some hundreds of trillions of molecules in a cubic inch of gas, and that these collide with each other eight thousand millions of times in a second. The average length of the path between two collisions of a molecule is less than the two hundred thousandth of an inch; yet this small length is supposed to be at least a hundred times as great as the diameter of each molecule.

From the fact that all gases expand with heat and contract with cold, it is concluded that the ether-vibrations we term heat are the cause of the rapid motions of the gaseous molecules, and that if heat was entirely absent the motion would cease, and ordinary cohesive attraction coming into play, the molecules would fall together and form a liquid or a solid. As a matter of fact, by intense cold, combined with pressure, all gases can be liquefied or solidified; and as, on the other hand, all the solid elements can be liquefied or vaporised by the intense heat of the electric furnace, we conclude that all matter when entirely deprived of heat is solid, and with sufficient heat becomes gaseous.

As might be expected from these varied phenomena, it has been found that there is no such sharp line of distinction between the various states of matter as is popularly supposed, some of the properties which are characteristic of matter in one state being present in a less degree in other states. Viscous bodies, for example, often present phenomena characteristic of both solids and fluids. Sealing-wax, pitch, and ice are all brittle at low temperatures, resembling in this respect such solids as glass and stone; but they are at the very same time fluid, if time enough is allowed to exhibit the phenomenon. This is seen in the motion of glaciers, which move in every respect like true fluids, even to the middle of the stream flowing quicker than the sides and the top than the bottom. Eddies and whirls occur in glaciers as in rivers, and also upward and downward motion, so that rccks torn off the glacier floor may be carried upwards and deposited on surfaces hundreds of feet above their place of origin. These properties can be shown to exist by experiment even on a small scale. A slab of ice, supported on its two ends, will become gradually curved, and the curvature may be increased to any desired extent if force is applied for a sufficient time. Models of glaciers in cobbler's wax, which is brittle at ordinary temperatures, exhibit all the phenomena of true glacier-motion, and serve to demonstrate the upward motions above referred to, which have been so often denied. Most metals exhibit similar phenomena under suitable conditions, and lead can be made actually to flow out of a hole under pressure.

One of the most characteristic properties of gases and liquids is that of readily mixing together when placed in contact. But it has recently been shown that solids also mix, though very much more slowly. If a cube of lead is placed upon one of gold, the surfaces of contact being very smooth and true, and be left without any pressure but their own weight, and at ordinary temperatures, for about a month, a minute quantity of gold will be found to have permeated through the lead, and can be detected in any part of it. Metals may thus be said to flow into each other.

In order to produce chemical changes in bodies, it is usually necessary that one at least be a liquid or be in a state of solution, and the combinations that occur lead to the production of bodies having quite different properties from either of their components. Similar results occur when metals are mixed together, forming alloys. Thus a mixture in certain proportions of lead, tin, bismuth, and cadmium produces an alloy which melts in boiling water, while the component metals only melt at double that temperature or more. Again, the strength of gold is doubled by the addition of one five-hundredth part of the rare metal zirconium, indicating that the alloy must have a new arrangement of the molecules. But the interesting point is, that alloys can be produced without melting the metals, for mere pressure often produces an alloy at the surfaces of contact; while in other cases if fine filings of the component metals are thoroughly mixed together and then subjected to continued pressure, true alloys are produced.

Another interesting fact is that metals, and probably all solids, evaporate at ordinary temperatures. It has long been known that ice evaporates very rapidly, and now it is found that metals do the same, and the evaporation can be detected at temperatures far below their melting points. All these curious phenomena give us new ideas as to the constitution of matter, and lead us to the conclusion that the extreme mobility of the molecules of gases has its analogue in liquids and even in solids. The flow of metals, their diffusion into other metals, and their evaporation, lead to the conclusion that a proportion of their molecules must possess considerable mobility, and when these reach the surface they are enabled to escape either into other bodies in contact with them or into the atmosphere. This proportion of rapidly moving molecules gives to solids some of the characteristics of liquids and of gases.

Before leaving this part of our subject we must refer to a most interesting and suggestive discovery which throws still further light on the constitution of matter, and on the forces which give to matter many of the properties without which neither vegetable nor animal life would be possible. It has been found that all gases expand or contract equal amounts for every degree of heat, the amount being $\frac{1}{273}$ of their volume for each degree centigrade. Hence, beginning at zero, if a gas could be cooled continuously down to -273 C, or -461 Fahr., it would not only be reduced to a solid, but would cease to have the power of further contraction. Hence this point is termed the absolute zero of temperature, and Lord Kelvin has arrived at the same result by quite different means. With the total absence of heat it is believed that all chemical action would cease, so that the universe would consist wholly of solid and chemically inert matter. Heat, therefore, seems to be the source of all change in matter, and the essential condition of all life; while the other vibrations of the ether, which we know as light and electricity, may be also essential. Ether, therefore, appears to be the active, matter the passive agent in the constitution of the universe; and the recognition of the existence of the ether, together with the considerable amount of knowledge we have acquired of its modes of action, must be held to constitute one of the most important intellectual triumphs of the Nineteenth Century.

# CHAPTER VIII

## MINOR APPLICATIONS OF PHYSICAL PRINCIPLES

Yes, thou shalt mark, with magic art profound,
The speed of light, the circling march of sound.

—*Campbell.*

O matchless Age! that even the passing tone
Of epoch-making speech, or lover's sigh,
Recordest for the wonder of all time!

—*F. T. Mott.*

AMONG the very numerous discoveries depending upon physical principles, or on the application of physical laws, a few of the more generally interesting may be here noticed.

The Radiometer, to be seen in almost every optician's window, was invented by Sir William Crookes in 1873, and consists of an exceedingly delicate windmill, formed of four very slender arms supporting thin metal or pith discs, one side of which is blackened, the whole turning on a fine central point, so as to revolve with hardly any friction. The little machine is enclosed in a glass bulb from which nearly all the air has been extracted; and when exposed to the sun, or even to diffused daylight, the discs revolve with considerable speed. At first this motion was supposed to be caused by the direct impact of the rays of light, the almost complete vacuum only serving to diminish friction; but the explanation now generally adopted is, that the black surfaces of the vanes, absorbing heat, become slightly warmer than the white surfaces, and this greater warmth is communicated to the air-molecules, and causes them to rebound with greater rapidity from the dark surfaces, and back again from the glass of the vessel, and the reaction being all in one direction, causes the arms to revolve. The near approach to a vacuum is necessary, both to diminish resistance, and by greatly reducing the number of molecules, in the vessel, to allow the very small differential action to produce a sensible effect. Sir William Crookes has found that there is a degree of rarefaction where the action is at a maximum, and that when a nearer approach to a perfect vacuum is attained the motion rapidly diminishes. A proof is thus given of the correctness of the explanation; and the instrument may, therefore, be considered to afford us an experimental illustration of the molecular theory of gases.

The velocity of light, as is well known, was first determined by irregularities in the time of the eclipses of Jupiter's satellites, which were found to occur earlier or later than the calculated times, according as we were near to, or far from, the planet. It was thus found that it required eight minutes for light to travel from the sun to the earth, a distance of a little more than ninety millions of miles; so that light travels about 196,000 miles in a single second of time. It would seem at first sight impossible to measure the time taken by light in travelling a mile, yet means have been discovered to do this, and even to measure the time taken for light to traverse a few feet from one side of a room to the other. Yet more, this method of measuring the velocity of light has, by successive refinements, become so accurate that it is now considered to be the most satisfactory method of determining the mean distance of the sun from the earth, a distance which serves as the unit of measurement for the solar system and the whole stellar universe. A brief account of how this is effected will now be given.

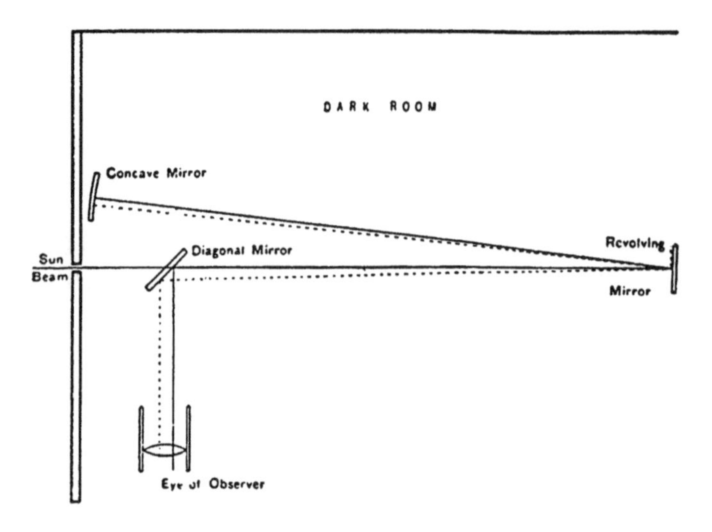


Fizeau, a French physicist, made the first attempt at measuring the velocity of light in 1849; and later, in 1862, in conjunction with Foucault, a more accurate determination was made by means of an apparatus of which the main features are given in the accompanying diagram. A ray of sunlight is made to enter a darkened room by a narrow slit, and falls on a mirror at the further side of the room, which can be made to revolve with great rapidity. From this it is reflected to a concave mirror having its centre of curvature at the revolving mirror. The diagonal mirror is transparent glass, through which the ray passes on its way to the revolving mirror, but on coming back a portion of the light is reflected at right angles to the eye of the observer. This involves much loss of light, and in more recent experiments the revolving mirror is slightly tilted, so that the returning ray passes beneath the outgoing ray, and is then reflected by a mirror or total reflexion prism to the eye of the observer. Now let us suppose the revolving mirror to be at rest. The various mirrors are first accurately adjusted, so that the narrow slit of light (or a fine wire in its centre) is so reflected by the three mirrors that it can be seen in the observing eye piece, and its position on a fine micrometer exactly noted. If now the mirror is slowly revolved the line of light will appear and disappear at each revolution; but if it is made to revolve more than thirty times a second, the line of light will be seen motionless, on the same principle that a rapidly moving luminous object is seen as an illuminated riband. But if light requires any time, however minute, to travel from the revolving to the concave mirror and back again, the mirror will during that time have turned a little on its axis, and the returning ray of light will be reflected to a slightly different point on the diagonal mirror and on the micrometer scale of the eye piece. In Foucault's experiment the distance between the concave and revolving mirrors was only thirteen and a half feet, and he had to make the mirror revolve six hundred times in a second before the returning ray was shifted rather less than one hundredth of an inch. By increasing the speed to eight hundred revolutions the distance was increased to about twelve thousandths of an inch, which, under a powerful magnifier, could be measured with great precision. Having measured with great accuracy the distance between the mirrors, and knowing the exact number of revolutions a second of the mirror, which was shown by a simple clockwork connected with it, the velocity of light was deduced as being 185,157 miles per second.

It is evident that there are here several sources of error. The short distance traversed by the light renders it necessary for the revolving mirror to turn with extreme rapidity, while the observed displacement of the ray is very small. Minute errors in the various measurements will therefore be enormously multiplied in the result. To obviate these difficulties the concave mirror has been placed much further away; and in the most recent and most accurate experiments by Professor Newcombe at Washington, the distance between the revolving and the concave mirrors was about two and a half miles, and the mirror revolved two hundred and thirty times a second. This gave such a large displacement of the returning ray that it could be measured with extreme accuracy, and the average of numerous trials gave the velocity of light as 186,327 miles per second. It thus appears that Foucault's measurement in a small room was only in error about $\frac{1}{180}$, or a little more than a half of one per cent, a wonderful testimony to his skill as an experimenter under such unfavourable conditions. Professor Newcombe believes that his determination is correct within $\frac{1}{10000}$, but he thinks that by placing the mirrors twenty or thirty miles apart in the clear atmosphere of the Rocky Mountains a still greater approach to perfect accuracy could be obtained.[1]

[1] For a more detailed account of Prof. Newcombe's experiments, see *Nature*, vol. xxxiv. p. 170.

The same M. Leon Foucault who made these beautiful experiments on the measurement of the velocity of light has also discovered a method by which the rotation of the earth on its axis can be experimentally demonstrated. When a heavy body is in free motion in any direction, it requires force to change the direction; and if no such force is applied, it will continue its motion in the same straight line or in the same plane. If a heavy pendulum is suspended from the axis of a horizontal wheel by a very long, thin wire, and if, when swinging in a fixed line across the room, the wheel is slowly turned, either the wire will twist a little or the ball forming the weight of the pendulum will revolve, but the plane in which the weight swings will not be altered. On the same principle, any pendulum freely swinging near the North Pole will not change the direction of its swing, although its point of support revolves in twenty-four hours with the earth's surface to which it is attached. On trying the experiment with a heavy weight suspended from the dome of the Pantheon in Paris and carefully set swinging, the plane of oscillation of the weight was found apparently to change at a uniform rate, and always in the same direction, which was opposite to that of the earth's rotation; proving that the surface of the earth moved round while the plane of oscillation remained fixed in space. This experiment can be tried in any place free from currents of air, such as a cellar. It only requires a heavy weight, say of 28 lbs., to be suspended by a string just strong enough to bear it. The weight must be drawn three or four feet away from the vertical line and fastened by a thread, so as to be set swinging by burning the thread without giving it any lateral motion. In an hour the line of swing will be found to have changed considerably, and in a direction opposite to that of the earth's rotation. At the North Pole a circle drawn on the surface turns completely round in twenty-four hours, so that a pendulum swung there with a circle beneath it divided like a twenty-four hour clock dial, would appear to revolve, and would tell the time. At the Equator, however, a circle on the surface of the earth does not itself rotate on its centre as at the Pole, but is merely carried round the earth with the north and south points of the circumference preserving the same direction in space. Therefore, at the Equator a pendulum should show no motion of rotation. At all intervening points it will appear to rotate, but slower and slower as we recede from the Pole; and mathematical calculation shows that, while at the Pole it apparently moves through an angle of 15° in an hour, at London it would move a little less than 12°, at Paris 11½°, at New York 9¾°, and at Ceylon somewhat less than 2° an hour. Experiments have been tried at each of these places, and the rate of apparent rotation of the pendulum has been found to agree very closely with the calculated amount, thus giving a complete proof that the apparent rotation is really due to the rotation to the earth on its axis. This mode of rendering the earth's rotation visible, in such a simple and convincing manner, is a discovery of considerable interest, even among the many wonderful discoveries of the century.

One more of these minor applications of scientific principles, leading to very startling results, must be briefly described. All sounds, including the infinitely varied modulations of the human voice, have long been known to be due to successive air-waves set up by various vibrating substances; but it would seem impossible by any mechanical means to reproduce these complex vibrations so exactly as to cause the words of the original speaker to be again heard, quite intelligibly, and with all their tones and modulations, at any distant time or place. Yet this has been done by means of the instrument called the phonograph, one of the many ingenious inventions of the American, Edison.

In the telephone this is effected instantaneously, through the medium of an electric current, which reproduces the vibrations set up by the voice of the speaker in a delicate elastic diaphragm by means of another diaphragm at the end of the conducting wire, perhaps hundreds of miles away, as already explained in Chapter III. In the phonograph the whole operation is mechanical. A diaphragm is set vibrating by the voice as in the telephone, but instead of being reproduced at a distance by means of an electric current, it registers itself permanently on a cylinder of very hard wax, as an indented spiral line. This is effected by means of a fine steel point, like a graving tool, connected by a delicate lever with the centre of the diaphragm. The wax cylinder turns and travels onward at a perfectly uniform rate, which can be delicately adjusted, so that the steel point if stationary will cut in it a very fine spiral groove, uniform in depth from end to end, the turns of the groove being very close to each other. But when the diaphragm is set vibrating by the voice of the speaker, the steel point moves rapidly up and down, and the resulting groove continually varies in depth, forming a complex series of undulations. If, now, the cylinder is shifted back so that the steel point is exactly where it was at starting, and the cylinder is then made to revolve and move onward at exactly the same rate as before, the up and down motions of the style, due to the irregular depth of the groove, set up the very same series of vibrations in the diaphragm as those which cut the groove; and these vibrations reproduce the voice with marvellous fidelity, so that the most complex and rapid speech, or the most exquisite singing, can be heard quite intelligibly, and with all their modulations and expressiveness, though not in exactly the same tone of voice.

The cylinders thus produced can be preserved for years, can be carried to any part of the world, and by means of a duplicate of the original instrument will there reproduce the words and the vocal peculiarities of the speaker. Phonographs are now largely manufactured, and are used for a variety of purposes. They serve for the rapid dictation of correspondence, which can be reproduced and copied by a clerk later on; to take down discussions verbatim, with a perfection that no shorthand writer can rival; the singing or the elocution of celebrated performers, is repeated for the gratification of friends or to amuse private parties; actors, musicians, and clergymen use the instrument as a means of improving their style; and even the languages, songs, and folk-lore of dying-out tribes are being preserved on these wonderful cylinders.

Probably there is no instrument in the world which so impresses the observer with the apparent inadequacy of the means to bring about so marvellous a result. At the same time it renders more mysterious than ever the properties and possibilities of air-waves, and the extreme delicacy of the ear and auditory nerves, which enable us instantaneously to interpret any one set of these vibrations, amidst the many other sets of air-waves arising from various sources which must be continually crossing and intermingling in apparently inextricable confusion. The phonograph, whether as illustrating the ingenuity of man or the marvellous perfection of our organism, will certainly take high rank among the new inventions of the nineteenth century.

# CHAPTER IX

## THE IMPORTANCE OF DUST: A SOURCE OF BEAUTY AND ESSENTIAL TO LIFE

When the lamp is shattered,
  The light in the dust lies dead;
When the cloud is scattered,
  The rainbow's glory is shed.

—*Shelley.*

How beautiful is the rain!
After the dust and heat,
In the broad and fiery street,
In the narrow lane,
How beautiful is the rain!

—*Longfellow.*

THE majority of persons, if asked what were the uses of dust, would reply that they did not know it had any, but they were sure it was a great nuisance. It is true that dust, in our towns and in our houses, is often not only a nuisance but a serious source of disease; while in many countries it produces ophthalmia, often resulting in total blindness. Dust, however, as it is usually perceived by us, is, like dirt, only matter in the wrong place, and whatever injurious or disagreeable effects it produces are largely due to our own dealings with nature. So soon as we dispense with horse-power and adopt purely mechanical means of traction and conveyance, we can almost wholly abolish disease-bearing dust from our streets, and ultimately from all our highways; while another kind of dust, that caused by the imperfect combustion of coal, may be got rid of with equal facility so soon as we consider pure air, sunlight, and natural beauty, to be of more importance to the population as a whole than are the prejudices or the vested interests of those who produce the smoke.

But though we can thus minimise the dangers and the inconveniences arising from the grosser forms of dust, we cannot wholly abolish it; and it is, indeed, fortunate we cannot do so, since it has now been discovered that it is to the presence of dust we owe much of the beauty, and perhaps even the very habitability, of the earth we live upon. Few of the fairy tales of science are more marvellous than these recent discoveries as to the varied effects and important uses of dust in the economy of nature.

The question why the sky and the deep ocean are both blue did not much concern the earlier physicists. It was thought to be the natural colour of pure air and water, so pale as not to be visible when small quantities were seen, and only exhibiting its true tint when we looked through great depths of atmosphere or of oceanic water. But this theory did not explain the familiar facts of the gorgeous tints seen at sunset and sunrise, not only in the atmosphere and on the clouds near the horizon, but also in equally resplendent hues when the invisible sun shines upon Alpine peaks and snowfields. A true theory should explain all these colours, which comprise almost every tint of the rainbow.

The explanation was found through experiments on the visibility or non-visibility of air, which were made by the late Professor Tyndall about the year 1868. Every one has seen the floating dust in a sunbeam when sunshine enters a partially darkened room; but it is not generally known that if there was absolutely no dust in the air the path of the sunbeam would be totally black and invisible, while if only very little dust was present in very minute particles the air would be as blue as a summer sky.

This was proved by passing a ray of electric light lengthways through a long glass cylinder filled with air of varying degrees of purity as regards dust. In the air of an ordinary room, however clean and well ventilated, the interior of the cylinder appears brilliantly illuminated. But if the cylinder is exhausted and then filled with air which has passed slowly through a fine gauze of intensely heated platinum wire, so as to burn up all the floating dust particles, which are mainly organic, the light will pass through the cylinder without illuminating the interior, which, viewed laterally, will appear as if filled with a dense black cloud. If, now, more air is passed into the cylinder through the heated gauze, but so rapidly that the dust particles are not wholly consumed, a slight blue haze will begin to appear, which will gradually become a pure blue, equal to that of a summer sky. If more and more dust particles are allowed to enter, the blue becomes paler, and gradually changes to the colourless illumination of the ordinary air.

The explanation of these phenomena is, that the number of dust-particles in ordinary air is so great that they reflect abundance of light of all wave-lengths, and thus cause the interior of the vessel containing them to appear illuminated with white light. The air which has passed slowly over white-hot platinum has had the dust particles destroyed, thus showing that they were almost wholly of organic origin, which is also indicated by their extreme lightness, causing them to float permanently in the atmosphere. The dust being thus got rid of, and pure air being entirely transparent, there is nothing in the cylinder to reflect the light which is sent through its centre in a beam of parallel rays, so that none of it strikes against the sides, hence the inside of the cylinder appears absolutely dark. But when all the larger dust particles are wholly or partially burnt, so that only the very smallest fragments remain, a blue light appears, because these are so minute as to reflect chiefly the more refrangible rays, which are of shorter wave-length—those at the blue end of the spectrum, which are thus scattered in all directions, while the red and yellow rays pass straight on as before.

We have seen that the air near the earth's surface is full of rather coarse particles which reflect all the rays, and which therefore produce no one colour. But higher up the particles necessarily become smaller and smaller, since the comparatively rare atmosphere will only support the very smallest and lightest. These exist throughout a great thickness of air, perhaps from one mile to ten miles high or even more, and blue or violet rays being reflected from the innumerable particles in this great mass of air, which is nearly uniform in all parts of the world as regards the presence of minute dust-particles, produces the constant and nearly uniform tint we call sky-blue. A certain amount of white or yellow light is no doubt reflected from the coarser dust in the lower atmosphere, and slightly dilutes the blue and renders it not quite so deep and pure as it otherwise would be. This is shown by the increasing depth of the sky-colour when seen from the tops of lofty mountains, while from the still greater heights attained in balloons the sky appears of a blue-black colour, the blue reflected from the comparatively small amount of dust particles being seen against the intense black of stellar space. It is for the same reason that the "Italian skies" are of so rich a blue, because the Mediterranean sea on one side and the snowy Alps on the other do not furnish so large a quantity of atmospheric dust in the lower strata of air as in less favourably situated countries, thus leaving the blue reflected by the more uniformly distributed fine dust of the higher strata undiluted. But these Mediterranean skies are surpassed by those of the central Pacific ocean, where, owing to the small area of land, the lower atmosphere is more free from coarse dust than any other part of the world.

If we look at the sky on a perfectly fine summer's day, we shall find that the blue colour is the most pure and intense overhead, and when looking high up in a direction opposite to the sun. Near the horizon it is always less bright, while in the region immediately round the sun it is more or less yellow. The reason of this is, that near the horizon we look through a very great thickness of the lower atmosphere, which is full of the larger dust-particles reflecting white light, and this dilutes the pure blue of the higher atmosphere seen beyond. And in the vicinity of the sun a good deal of the blue light is reflected back into space by the finer dust, thus giving a yellowish tinge to that which reaches us reflected chiefly from the coarse dust of the lower atmosphere.

At sunset and sunrise, however, this last effect is greatly intensified, owing to the great thickness of the strata of air through which the light reaches us. The enormous amount of this dust is well shown by the fact that, then only, we can look full at the sun, even when the whole sky is free from clouds and there is no apparent mist. But the sun's rays then reach us after having passed, first, through an enormous thickness of the higher strata of the air, the minute dust of which reflects most of the blue rays away from us, leaving the complementary yellow light to pass on. Then, the somewhat coarser dust reflects the green rays, leaving a more orange coloured light to pass on; and finally some of the yellow is reflected, leaving almost pure red. But owing to the constant presence of air currents, arranging both the dust and vapour in strata of varying extent and density, and of high or low clouds, which both absorb and reflect the light in varying degrees, we see produced all those wondrous combinations of tints and those gorgeous ever-changing colours, which are a constant source of admiration and delight to all who have the advantage of an uninterrupted view to the west, and who are accustomed to watch for these not unfrequent exhibitions of nature's kaleidoscopic colour-painting. With every change in the altitude of the sun the display changes its character; and most of all when it has sunk below the horizon, and, owing to the more favourable angles, a larger quantity of the coloured light is reflected towards us. Especially when there is a certain amount of cloud is this the case. These, so long as the sun was above the horizon, intercepted much of the light and colour; but, when the great luminary has passed away from our direct vision, his light shines more directly on the under sides of all the clouds and air strata of different densities; a new and more brilliant light flushes the western sky, and a display of gorgeous ever-changing tints occurs which are at once the delight of the beholder and the despair of the artist. And all this unsurpassable glory we owe to—dust!

A remarkable confirmation of this theory was given during the two or three years after the great eruption of Krakatoa, near Java. The volcanic *débris* was shot up from the crater many miles high, and the heavier portion of it fell upon the sea for several hundred miles around, and was found to be mainly composed of very thin flakes of volcanic glass. Much of this was of course ground to impalpable dust by the violence of the discharge, and was carried up to a height of many miles. Here it was caught by the return current of air continually flowing northwards and southwards above the equatorial zone; and as these currents reach the temperate zone where the surface rotation of the earth is less rapid they continually flow eastward, and the fine dust was thus carried at a great altitude completely round the earth. Its effects were traced some months after the eruption in the appearance of brilliant sunset glows of an exceptional character, often flushing with crimson the whole western half of the visible sky. These glows continued in diminishing splendour for about three years; they were seen all over the temperate zone, and it was calculated that, before they finally disappeared, some of this fine dust must have travelled three times round the globe.

The same principle is thought to explain the exquisite blue colour of the deep seas and oceans and of many lakes and springs. Absolutely pure water, like pure air, is colourless, but all seas and lakes, however clear and translucent, contain abundance of very finely divided matter, organic or inorganic, which, as in the atmosphere, reflects the blue rays in such quantity as to overpower the white or coloured light reflected from the fewer and more rapidly sinking particles of larger size. The oceanic dust is derived from many sources. Minute organisms are constantly dying near the surface, and their skeletons, or fragments of them, fall slowly to the bottom. The mud brought down by rivers, though it cannot be traced on the ocean floor more than about 150 miles from land, yet no doubt furnishes many particles of organic matter which are carried by surface currents to enormous distances and are ultimately dissolved before they reach the bottom. A more important scource of finely divided matter is to be found in volcanic dust which, as in the case of Krakatoa, may remain for years in the atmosphere, but which must ultimately fall upon the surface of the earth and ocean. This can be traced in all the deep sea oozes. Finally there is meteoric dust, which is continually falling to the surface of the earth, but in such minute quantities and in such a finely-divided state that it can only be detected in the oozes of the deepest oceans where both inorganic and organic *debris* is almost absent.

The blue of the ocean varies in different parts from a pure blue somewhat lighter than that of the sky, as seen about the northern tropic in the Atlantic, to a deep indigo tint, as seen in the north temperate portions of the same ocean: due, probably, to differences in the nature, quantity, and distribution of the solid matter which causes the colour. The Mediterranean, and the deeper Swiss lakes are also blue of various tints, due also to the presence of suspended matter, which Professor Tyndall thought might be so fine that it would require ages of quiet subsidence to reach the bottom. All the evidence goes to show, therefore, that the exquisite blue tints of sky and ocean, as well as all the sunset hues of sky and cloud, of mountain peak and alpine snows, are due to the finer particles of that very dust which, in its coarser forms, we find so annoying and even dangerous.

But if this production of colour and beauty were the only useful function of dust, some persons might be disposed to dispense with it in order to escape its less agreeable effects. It has, however, been recently discovered that dust has another part to play in nature, a part so important that it is doubtful whether we could even live without it. To the presence of dust in the higher atmosphere we owe the formation of mists, clouds, and gentle beneficial rains, instead of waterspouts and destructive torrents.

It is barely twenty years ago since the discovery was made, first in France by Coulier and Mascart, but more thoroughly worked out by Mr. John Aitken in 1880. He found that if a jet of steam is admitted into two large glass receivers, one filled with ordinary air the other with air which has been filtered through cotton wool so as to keep back all particles of solid matter, the first will be instantly filled with condensed vapour in the usual cloudy form, while the other vessel will remain quite transparent. Another experiment was made more nearly reproducing what occurs in nature. Some water was placed in the two vessels prepared as before. When the water had evaporated sufficiently to saturate the air the vessels were slightly cooled, when a dense cloud was at once formed in the one while the other remained quite clear. These experiments and many others, showed, that the mere cooling of vapour in air will not condense it into mist clouds or rain, unless *particles of solid matter* are present to form *nuclei* upon which condensation can begin. The density of the cloud is proportionate to the number of the particles; hence the fact that the steam issuing from the safety-valve or the chimmey of a locomotive forms a dense white cloud, shows that the air is really full of dust-particles, most of which are microscopic but none the less serving as centres of condensation for the vapour. Hence, if there were no dust in the air, escaping steam would remain invisible; there would be no clouds in the sky; and the vapour in the atmosphere, constantly accumulating through evaporation from seas and oceans and from the earth's surface, would have to find some other means of returning to its source.

One of these modes would be the deposition of dew, which is itself an illustration of the principle that vapour requires solid or liquid surfaces to condense upon; hence dew forms more readily and more abundantly on grass, on account of the numerous centres of condensation it affords. Dew, however, is now formed only on clear cold nights after warm or moist days. The air near the surface is warm and contains much vapour, though below the point of saturation. But the innumerable points and extensive surfaces of grass radiate heat quickly, and becoming cool, lower the temperature of the adjacent air, which then reaches saturation point and condenses the contained vapour on the grass. Hence, if the atmosphere at the earth's surface became super-saturated with aqueous vapour, dew would be continuously deposited, especially on every form of vegetation, the result being that everything, including our clothing, would be constantly dripping wet. If there were absolutely no particles of solid matter in the upper atmosphere, all the moisture would be returned to the earth in the form of dense mists, and frequent and copious dews, which in forests would form torrents of rain by the rapid condensation on the leaves. But if we suppose that solid particles were occasionally carried higher up through violent winds or tornadoes, then on those occasions the super-saturated atmosphere would condense rapidly upon them, and while falling would gather almost all the moisture in the atmosphere in that locality, resulting in masses or sheets of water, which would be so ruinously destructive by the mere weight and impetus of their fall that it is doubtful whether they would not render the earth almost wholly uninhabitable.

The chief mode of discharging the atmospheric vapour in the absence of dust, would, however, be by contact with the higher slopes of all mountain ranges. Atmospheric vapour being lighter than air, would accumulate in enormous quantities in the upper strata of the atmosphere, which would be always super-saturated and ready to condense upon any solid or liquid surfaces. But the quantity of land comprised in the upper half of all the mountains of the world is a very small fraction of the total surface of the globe, and this would lead to very disastrous results. The air in contact with the higher mountain slopes would rapidly discharge its water, which would run down the mountain sides in torrents. This condensation on every side of the mountains would leave a partial vacuum, which would set up currents from every direction to restore the equilibrium, thus bringing in more super-saturated air to suffer condensation and add its supply of water, again increasing the in-draught of more air. The result would be, that winds would be constantly blowing towards every mountain range from all directions, keeping up the condensation and discharging, day and night and from one year's end to another, an amount of water equal to that which falls during the heaviest tropical rains. The whole of the rain that now falls over the whole surface of the earth and ocean with the exception of a few desert areas, would then fall only on rather high mountains or steep isolated hills, tearing down their sides in huge torrents, cutting deep ravines, and rendering all growth of vegetation impossible. The mountains would therefore be so devastated as to be uninhabitable, and would be equally incapable of supporting either vegetable or animal life.

But this constant condensation on the mountains would probably check the deposit on the lowlands in the form of dew, because the continual up-draught towards the higher slopes would withdraw almost the whole of the vapour as it rose from the oceans and other water-surfaces, and thus leave the lower strata over the plains almost or quite dry. And if this were the case there would be no vegetation, and therefore no animal life, on the plains and lowlands, which would thus be all arid deserts cut through by the great rivers formed by the meeting together of the innumerable torrents from the mountains.

Now, although it may not be possible to determine

with prefect accuracy what would happen under the supposed condition of the atmosphere, it is certain that the total absence of dust would so fundamentally change the meteorology of our globe as, not improbably, to render it uninhabitable by man, and equally unsuitable for the larger portion of its existing animal and vegetable life.

Let us now briefly summarize what we owe to the universality of dust, and especially to that most finely divided portion of it which is constantly present in the atmosphere up to the height of many miles. First of all it gives us the pure blue of the sky, one of the most exquisitely beautiful colours in nature. It gives us also the glories of the sunset and the sunrise, and all those brilliant hues seen in high mountain regions. Half the beauty of the world would vanish with the absence of dust. But, what is far more important than the colour of sky and beauty of sunset, dust gives us also diffused daylight, or skylight, that most equable, and soothing, and useful, of all illuminating agencies. Without dust the sky would appear absolutely black, and the stars would be visible even at noonday. The sky itself would therefore give us no light. We should have bright glaring sunlight or intensely dark shadows, with hardly any half-tones. From this cause alone the world would be so totally different from what it is, that all vegetable and animal life would probably have developed into very different forms, and even our own organisation would have been modified in order that we might enjoy life in a world of such harsh and violent contrasts.

In our houses we should have little light except when the sun shone directly into them, and even then every spot out of its direct rays would be completely dark, except for light reflected from the walls. It would be necessary to have windows all round and the walls all white; and on the north side of every house a high white wall would have to be built to reflect the light and prevent that side from being in total darkness. Even then we should have to live in a perpetual glare, or shut out the sun altogether and use artificial light as being a far superior article.

Much more important would be the effects of a dust-free atmosphere in banishing clouds, or mist, or the "gentle rain of heaven," and in giving us in their place perpetual sunshine, desert lowlands, and mountains devastated by unceasing floods and raging torrents, so as, apparently, to render all life on the earth impossible.

There are a few other phenomena, apparently due to the same general causes, which may here be referred to. Every one must have noticed the difference in the atmospheric effects and general character of the light in spring and autumn, at times when the days are of the same length, and consequently when the sun has the same altitude at corresponding hours. In spring we have a bluer sky and greater transparency of the atmosphere; in autumn, even on very fine days, there is always a kind of yellowish haze, resulting in a want of clearness in the air and purity of colour in the sky. These phenomena are quite intelligible when we consider that during winter less dust is formed, and more is brought down to the earth by rain and snow, resulting in the transparent atmosphere of spring, while exactly opposite conditions during summer bring about the mellow autumnal light. Again, the well known beneficial effects of rain on vegetation, as compared with any amount of artificial watering, though, no doubt, largely due to the minute quantity of ammonia which the rain brings down with it from the air, must yet be partly derived from the organic or mineral particles which serve as the nuclei of every rain drop, and which, being so minute, are more readily dissolved in the soil and appropriated as nourishment by the roots of plants.

It will be observed that all these beneficial effects of dust are due to its presence in such quantities as are produced by natural causes, since both gentle showers as well as ample rains and deep blue skies, are present throughout the vast equatorial forest districts, where dust-forming agencies seem to be at a minimum. But in all densely-populated countries there is an enormous artificial production of dust—from our ploughed fields, from our roads and streets, where dust is continually formed by the iron-shod hoofs of innumerable horses, but chiefly from our enormous combustion of fuel pouring into the air volumes of smoke charged with unconsumed particles of carbon. This superabundance of dust, probably many times greater than that which would be produced under the more natural conditions which prevailed when our country was more thinly populated, must almost certainly produce some effect on our climate; and the particular effect it seems calculated to produce is the increase of cloud and fog, but not necessarily any increase of rain. Rain depends on the supply of aqueous vapour by evaporation; on temperature, which determines the dew point; and on changes in barometric pressure, which determine the winds. There is probably always and everywhere enough atmospheric dust to serve as centres of condensation at considerable altitudes, and thus to initiate rainfall when the other conditions are favourable; but the presence of increased quantities of dust at the lower levels must lead to the formation of denser clouds, although the minute water-vesicles cannot descend as rain, because, as they pass down into warmer and dryer strata of air, they are again evaporated.

Now, there is much evidence to show that there has been a considerable increase in the amount of cloud, and consequent decrease in the amount of sunshine, in all parts of our country. It is an undoubted fact that in the middle ages England was a wine-producing country, and this implies more sunshine than we have now. Sunshine has a double effect, in heating the surface soil and thus causing more rapid growth, besides its direct effect in ripening the fruit. This is well seen in Canada, where, notwithstanding a six months' winter of extreme severity, vines are grown as bushes in the open ground, and produce fruit equal to that of our ordinary greenhouses. Some years back one of our gardening periodicals obtained from gardeners of forty or fifty years' experience a body of facts clearly indicating a comparatively recent change of climate. It was stated that in many parts of the country, especially in the north, fruits were formerly grown successfully and of good quality in gardens where they cannot be grown now; and this occurred in places sufficiently removed from manufacturing centres to be unaffected by any direct deleterious influence of smoke. But an increase of cloud, and consequent diminution of sunshine, would produce just such a result; and this increase is almost certain to have occurred, owing to the enormously increased amount of dust thrown into the atmosphere as our country has become more densely populated, and especially owing to the vast increase of our smoke-producing manufactories. It seems highly probable, therefore, that to increase the wealth of capitalist-manufacturers we are allowing the climate of our whole country to be greatly deteriorated in a way which diminishes both its productiveness and its beauty, thus injuriously affecting the enjoyment and the health of the whole population, since sunshine is itself an essential condition of healthy life. When this fact is thoroughly realized we shall surely put a stop to such a reckless and wholly unnecessary production of injurious smoke and dust.

In conclusion, we find that the much-abused and all-pervading dust, which, when too freely produced, deteriorates our climate and brings us dirt, discomfort, and even disease, is, nevertheless, under natural conditions, an essential portion of the economy of nature. It gives us much of the beauty of natural scenery as due to varying atmospheric effects of sky, and cloud, and sunset tints, and thus renders life more enjoyable; while, as an essential condition of diffused daylight, and of moderate rainfalls combined with a dry atmosphere, it appears to be absolutely necessary for our existence upon the earth, perhaps even for the very development of terrestrial, as opposed to aquatic life. The overwhelming importance of the small things, and even of the despised things of our world, has never, perhaps, been so strikingly brought home to us as in these recent investigations into the wide-spread and far-reaching beneficial influences of Atmospheric Dust.

# CHAPTER X

## A FEW OF THE GREAT PROBLEMS OF CHEMISTRY

Force merges into force
  The atom seeks its kind;
The elements are one,
  And each with all combined.

—*F. T. Palgrave.*

O Lavoisier, master great,
We mourn your awful fate,
But never tire of singing to your praise.
You laid foundations true,
And we must trace to you
The chemistry of our enlightened days.

—*Anon.*

THE science of modern chemistry has been created during the present century, but its phenomena and laws are so complex that it presents only a few of those great discoveries which are the starting points for new developments, and which can at the same time be popularly described. The most important of all—that which constitutes the very foundation of chemistry as a science—is the law of chemical combination in multiple proportions, together with the atomic theory which serves to explain it.

The fact of chemical combination in definite proportions was suspected by some of the older chemists, but Dalton, in the early years of this century, was the first to establish it firmly as a general principle, and to explain it by means of a comparatively simple theory. To illustrate by examples, it is found that the two gases, nitrogen and oxygen, combine to form a variety of compounds, such as nitrous oxide or "laughing gas," nitric oxide, and several others. Nitrous oxide, or in chemical language, nitrogen monoxide, consists of 28 parts by weight of nitrogen to 16 of oxygen, and all the other compounds of the same gases consist of two, three, four, or five times as much oxygen to the same quantity of nitrogen. Water consists of 16 parts of oxygen to 2 of hydrogen, and there is another compound in which 32 parts of oxygen combine with the same weight of hydrogen, forming hydrogen-dioxide or oxygenated water. This law applies to every chemical compound yet discovered, and as every element has a minimum proportionate weight, which can combine with any other element, these are called the atomic or combining weights of the elements. As the weight of the hydrogen in all its combinations is much less than the weight of the element it combines with, this gas is taken as the unit of measurement of atomic weights. Nitrogen is thus found to have an atomic weight of 14, oxygen 16, and chlorine 35. These are all gases; but many solids have much lower atomic weights, carbon being 12, and the rare metal beryllium only 9. Of other metals, that of aluminium is 27, copper 63, iron 56, silver 107, tin 117, and gold 196. There is thus no constant relation between atomic weights and specific gravities. Tin is a little lighter than iron, but has nearly double its atomic weight; gold has a high atomic weight, but bismuth has a higher still, although only half its specific gravity.

These facts are elucidated, and to some extent explained, by the atomic theory of Dalton. He supposed each element to consist of atoms, an atom being the smallest portion that has the properties of the element, and the atom of each element has a different weight. Hence when one element combines with another the proportions must be either those represented by the atomic weights, or some multiple of those weights, since the atoms are assumed to be indivisible. This will be made clearer by another example. The atomic weights of nitrogen and oxygen are as 14 to 16, and these elements combine in five different proportions, as shown by the following figures, each circle representing an atom of the elements indicated by their initial letters:—

| | | Chemical Symbol |
|---|---|---|
| (N)(N)(O) | =Nitrogen monoxide | $N_2O$ |
| (N)(N)(O)(O) | =Nitrogen dioxide | $N_2O_2$ |
| (N)(N)(O)(O)(O) | =Nitrogen trioxide | $N_2O_3$ |
| (N)(N)(O)(O)(O)(O) | =Nitrogen tetroxide | $N_2O_4$ |
| (N)(N)(O)(O)(O)(O)(O) | =Nitrogen pentoxide | $N_2O_5$ |

The atomic or combining weights of all the elements having been carefully determined by numerous experiments, a beautiful system of chemical symbols has been formed which greatly facilitates the study of the innumerable complex substances that have to be investigated. Each element is indicated either by one or two letters, being the initial letter, or some two characteristic letters, of its chemical name, so that nearly seventy elements are thus clearly defined. But these symbols represent not only the element, but a definite proportional weight—the atomic weight. Thus H means a unit weight of hydrogen; C means 12 times that weight of carbon; Fe (ferrum) means 56 times that weight of iron. Hence the symbol for any compound substance tells us in the most compact form possible, not only the elements of which it is composed, but the exact proportions in which these elements are combined. Thus $C_2H_6O$ is the chemical symbol for pure alcohol, showing that it is a compound of two atoms of carbon, six of hydrogen, and one of oxygen. Looking now at a table of atomic weights, we find that this gives us 24 carbon, 6 hydrogen, and 16 oxygen in each 46 parts of alcohol. By means of these symbols and the accurate determination of atomic weights, all the complex combinations and decompositions that occur during the investigations of the chemist can be represented in a kind of chemical algebra, and the peculiar formulæ thus obtained often suggest further experiments leading to new discoveries.

Almost at the same time that Dalton was working at his atomic theory, Davy (afterwards Sir Humphry Davy) made the remarkable discovery of two new elements by decomposing soda and potash by means of an electric current, resulting in the production of the metals, sodium and potassium. This placed in the hands of chemists a powerful agent which led to the discovery of other elements, though in this respect it has been surpassed by spectrum-analysis, which is equally effective in the domains of chemistry and astronomy.

Among the more interesting discoveries of modern chemistry are the methods of liquefying the various gases, and even solidifying many of them; while by means of the intense heat of the electric furnace all the solid elements can be melted and many vaporized, leading to the conclusion that all matter can exist in the three states—solid, liquid, and gaseous,—according to the degree of heat to which it is exposed.

The highly complex constitution of various organic products—albumen, fat, gums, resins, acids, oils, ethers, etc.—is the subject of organic chemistry, the study of which has led to some of the most popularly interesting discoveries. Coal-tar has furnished us with a wonderful series of colouring matters, such as the aniline and other dyes, while from the same material are produced benzol, carbolic acid, naphtha, creosote, artificial quinine, and saccharine, a substitute for sugar. The new explosives, such as dynamite and nitro-glycerine, are produced from animal or vegetable fatty matters; while some of the greatest triumphs of the modern chemist are the artificial production of natural substances, which were long supposed to be due to organic processes alone. Such are the dye indigo, citric acid, urea, and some others.

The most recent great advance in the philosophy of chemistry is exhibited in the views of the Russian chemist, Mendeleef, as to the natural arrangement of the elements, with certain deductions from it. The whole of the best known elements form eight groups, placed in vertical columns, depending on certain similarities in their powers of chemical combination. These are further arranged in twelve horizontal series, in which the atomic weights are most nearly alike, while increasing regularly from the first to the eighth group. In the table thus formed there are certain gaps in the regular order of increase of atomic weights, as if some elements were wanting, while in other cases the place of an element due to its atomic weight did not accord with that dependent on its chemical properties. But the general symmetry of the whole arrangement was such that Mendeleef predicted the future discovery of elements to fill the gaps, and named the chemical and physical properties of these unknown elements. In a few years three new elements were discovered—gallium, scandium, and germanium—and they precisely filled up three of the gaps in the system. Further research as to the atomic weights of the elements that did not fit into the scheme showed that errors had been made, that of uranium being much too low, while in the cases of gold, tellurium, and titanium it was too great. The remarkable success of these predictions—a success always considered the best proof of the truth of a theory—renders it almost certain that the true relations of the elements have now been approximately ascertained, while it strengthens the belief of those who think that what we term elements are not really so, but that their differences depend on special modes of aggregation of a few simple atoms, whose cohesion is so strong that we are not yet, and perhaps never shall be, able to overcome it.

It is therefore by no means impossible, perhaps not even improbable, that methods will be discovered of either breaking up some of the elements and producing new elements which are common to two or more of them, or of solving the problem which occupied the alchemists of the middle ages—the transmutation of some of the inferior metals into gold. Within the last few months a well-known American chemist declares that he has solved the problem of producing gold out of silver at a comparatively small cost, and that when he has made a few millions by his process he will make it known. A few years ago this claim would have been scouted as that of a dreamer, but at the present day it is really less unexpected than was the discovery of the marvellous powers of what are termed the Röntgen rays.

It will thus be seen, that chemistry, as a science, has not furnished discoveries of such a startling nature as those in the domain of physics. But this is largely due to the fact that we have already, in our earlier chapters, dealt with the more popular and industrial aspects of chemical inventions. Gas illumination, petroleum oil-lamps, lucifer matches, and all the wonders of photography are essentially applications of chemistry; and the last of these, in its marvellous results, both in the arts and in its various applications to astronomical research, is not surpassed by the achievements of any other department of science.

# CHAPTER XI

## ASTRONOMY AND COSMIC THEORIES

The wilder'd mind is tost and lost,
  O sea, in thy eternal tide;
The reeling brain essays in vain,
  O stars, to grasp the vastness wide!
The terrible, tremendous scheme
  That glimmers in each glancing light,
O night, O stars, too rudely-jars
  The finite with the infinite!

—*J. H. Dell.*

MANY of the most striking discoveries in this science have been already described under Spectrum Analysis; but there remain a few great advances, due either to observation or to theory, which are of sufficient popular interest to demand notice in any sketch, however brief, of the scientific progress of the century.

With the single exception of Uranus, discovered by Herschell in 1781, no addition had been made to the five planets known to the ancients till the commencement of the present century, when Ceres, the first of the minor planets, was discovered in 1801, and three others between that date and 1807. No more were found till one was added in 1845, and another in 1847. Since that time no year has passed without the detection of one or more new planets belonging to the same system, till in September 1896, their number amounted to 417. These small bodies form a kind of planetary ring situated between Mars and Jupiter, where it had long been suspected a planet ought to be found, because the distance between these older planets was so great as to be quite out of proportion with the regular increase of distance maintained by the other members of the system. It was at first thought that these asteroids or minor planets were the shattered remains of a much larger one; but more extended knowledge of the constitution of the solar system renders it more probable that they really constitute a ring of matter thrown off by the sun during its progressive contraction; and that some peculiar conditions have prevented its various parts from aggregating into a single planet. This is rendered more probable by two other remarkable discoveries relating to meteors and comets, and to Saturn's rings, which will be discussed later on.

The next large planet added to our system is especially interesting, as affording a striking demonstration of the theory of gravitation, and a no less striking example of the powers of modern mathematics. It had been found that the motions of Uranus were not exactly what they ought to be, if due solely to the attraction of the sun and the disturbing influence of Jupiter and Saturn, and it was thought possible that there might be another planet beyond it to cause these irregularities. In the year 1843 a young Cambridge student (John Couch Adams) of the highest mathematical ability, determined to see whether it was not possible to prove the existence of such a planet; and having taken his degree as Senior Wrangler, he at once devoted himself to the work, and after two years of study and calculation he was able to declare that a planet which would account for the perturbations of Uranus must, if it existed, be at that time in a certain part of the heavens, and he sent his paper on the subject to the Astronomer-Royal in October, 1845. By an extraordinary coincidence, a French astronomer (Leverrier) had been for some years working out the motions of the various planets, and in doing so had also reached the conclusion that there must be another unknown body to produce the perturbations of Uranus, which were at that time unusually large. His calculations and results were published at Paris in November, 1845, and June, 1846, and he gave a position for the unknown planet differing only one degree from that given by Adams. On reading these papers, and seeing the agreement of two independent workers, the Astronomer-Royal asked Professor Challis, of the Cambridge Observatory, to search for the planet, and on doing so he actually observed it on August 4th, and again on August 12th; but having no accurate chart of that part of the heavens he could not be sure that it was not a small star. A month later it was found and identified at Berlin, from information furnished by Leverrier. It thus appears that Adams first privately announced the position of the new planet, and that it was first observed at Cambridge; while the somewhat later announcement by Leverrier and discovery at Berlin were made public, and thus gained the honours of priority. The two discoveries were, however, practically simultaneous and independent, and the names of Adams and Leverrier should for ever be jointly associated with the planet Neptune.

Other important discoveries in the planetary system are due to the increased power of modern telescopes and the greater number of observers. In 1877 two minute satellites of Mars were discovered at Washington, by means of the large telescope with a 25-inch object glass, then the largest in the world. These are remarkable in being exceedingly small, and very close to the planet. They are said to be only six or seven miles in diameter, and the inner one is only about 5,800 miles from the centre, or 3,800 from the surface, of the planet, around which it revolves in less than eight hours; while the outer one is about 14,500 miles away, and revolves in a little more than thirty hours.[1]

Still more recently (in September, 1892), a fifth satellite of Jupiter was discovered by means of the great Lick telescope in California. This also is very small and very close to the planet, being less than half the diameter, or about 40,000 miles, from its surface.

Another very remarkable discovery is that of a system of symmetrical markings, covering a large part of the surface of Mars. They consist of a series of triangles or quadrilaterals bounded by straight lines, which are sometimes seen double, at other times single, or are even altogether invisible. Another peculiar feature is, that where these canals (as they are termed) intersect there is always a black circular spot, very distinct, and unlike any markings upon other parts of the surface. It is a curious fact that the double canals sometimes enclose a space of more

[1] In *Gulliver's Travels*, published in 1726, Swift describes the astronomers of Laputa as having "discovered two lesser stars, or satellites, which revolve around Mars; whereof the innermost is distant from the centre of the primary planet exactly three of his diameters, and the outermost five; the former revolves in the space of ten hours, and the latter in twenty-one and a half." This is a wonderful anticipation, especially as to time of revolution, and if we substitute "radii" for "diameters," the distances are also very near.

than a hundred miles wide and several hundred long, adding to the appearance of artificiality. Sometimes no canals are seen, but they come into view as the polar snows begin to melt; hence the suggestion that they really indicate great canals to carry off the water from the rapidly-melting snow and distribute it by irrigation channels over the adjacent land, which, being rapidly covered with vegetation, causes the change of colour which renders them visible. These observations were made by Mr. Percival Lowell during the favourable opposition, in 1894, at his observatory in Arizona, where the exceptional purity of the atmosphere renders it possible almost constantly to observe details which are elsewhere rarely visible. If future observations should confirm the views as to the artificial nature of these features of the surface of the planet which most nearly resembles our earth, it must be considered to be the most sensational astronomical discovery of the nineteenth century, and that which opens up the most exciting possibilities as to communication with beings who are sufficiently advanced to execute such widespread and gigantic irrigation works.

### *Saturn's Rings, and the Meteoritic Theory of the Universe*

The ring around the planet Saturn was long supposed to be single, and to be solid like the planet itself; but with improved telescopes it was found to be double, and with still finer instruments to consist of an indefinite number of rings close together, one of them being very obscure, as if formed of nebulous matter. In the year 1859, Clerk-Maxwell, by a profound mathematical investigation, proved that either solid or liquid rings would be unstable, and would inevitably break up so as to form a number of satellites; and he concluded that the rings really consisted of a crowd of small bodies so near together as to appear like a solid mass; and as the appearance of the rings, and some slight changes detected in them, were in harmony with this view, it has been generally accepted. But quite recently the wonderful instrument, the spectroscope, has given the final demonstration that this theory is correct. If the rings are solid, it is clear that a point on the outer edge must move more rapidly than one on the inner edge; whereas, if they consist of separate particles, each revolving independently round the planet, then, in accordance with the laws of all planetary motions, those forming the inner side of the rings, being nearer to the planet, must move much quicker than those on the outer side. As already explained in Chapter VI., the spectroscope enables us to measure motion in the line of sight—that is, towards or away from us—of any heavenly bodies, and by observing the outer extremities of the rings to the right and left of the planet, where the motion is, of course, in these two directions, it is found that the motion of the inner edge is considerably more rapid than that of the outer edge, showing that those parts move round the planet independently, and are therefore formed of separate particles or small masses. These observations were made by the American astronomer, Professor James E. Keeler, in 1895, and are of extreme delicacy; but that they are trustworthy is shown by the fact that the resulting velocities are in accordance with Kepler's third law, which determines the relative motions of all planetary bodies at varying distances from the primary.

A still more important discovery is that which has explained, by one consistent theory, the various phenomena presented by aërolites, fireballs, and shooting or falling stars, now generally classed as meteors and meteorites; and this theory is found to have an important bearing on the constitution of the solar system, and perhaps even on that of the whole stellar universe. Although there are records of the fall of solid stones from the sky in the works of classical, Chinese, and European authors, from 654 B.C. down to our times, while the astronomer Gassendi himself witnessed the fall of a stone weighing 59 lbs. in the year 1627, in the south of France, yet the phenomenon was so rare, and so inexplicable, that it was often disbelieved. One philosopher is reported to have disposed of the whole matter by saying, "there are no stones in the sky, therefore none can fall from it." But the evidence for such falls soon became overwhelming, and their connection with fireballs and shooting stars was also well established. One of the most remarkable of modern meteors was that seen at Aigle in Normandy, on April 26th, 1803. About 1 p.m. a brilliant fireball was seen traversing the air at great speed. A violent explosion followed, apparently proceeding from a small lofty cloud. This was no doubt the product of the explosion which would become visible long before the sound was heard; and then came a perfect shower of stones, nearly three thousand being picked up, the largest weighing eight pounds. A still more extraordinary meteor was seen on March 19th, 1719, about eight o'clock in the evening, in all parts of England, Scotland, and Ireland. In London it appeared like a ball of fire as large as the moon; at Exeter the light was like that of the sun. It was followed by a broad stream of light, and burst with a report like that of a cannon, with a great display of red sparks like a huge sky-rocket; but as it was then over the sea, between Devonshire and the coast of Brittany, its fragments were not recoverable. Dr. Whiston, Newton's successor as professor of mathematics at Cambridge, who published an account of it, calculated its height over London as 51 miles, and over Devonshire 39 miles.

Falling stars, sometimes seen singly, at other times in considerable numbers, as well as the less frequent but larger fireballs above described, appeared to be connected phenomena, although little was really known about them till the early part of this century, when they began to be more carefully studied. By observations of the same meteor or fireball at distant localities, its altitude, and the velocity with which it moved, were ascertained, and these were always found to be so great as to show that these objects could not have a terrestrial origin. It was soon observed that showers of falling stars occurred about the same time every year, with displays of great brilliancy at long intervals; and on these occasions the meteors all appeared to radiate from certain definite points in the sky. Thus in November they seemed to originate in the constellation Leo, and in August in Perseus, while others apparently belong to distinct constellations. The only way of explaining these appearances seemed to be, that there were streams of small bodies travelling in elliptic orbits round the sun, and that the earth crossed these orbits at fixed points once a year. Then, a number of these small bodies, many of them perhaps no larger than pebbles or grains of sand, coming into our atmosphere, became heated and even vaporized by the friction due to their rapid planetary motion, and appeared to us as shooting stars; while larger masses, whose exterior alone became heated, either exploded or fell entire as meteorites. The exceptional displays of the November meteors at intervals of about thirty-three years is due to the fact that the stream is much denser in this part of its orbit, where the meteoric matter may be slowly aggregating to form a planetary body.

A large number of such meteor streams have now been observed; but the most remarkable discovery is, that in some cases, and probably in all, comets form a part of such meteoric streams. This has been proved by showing that the orbits and times of revolution of certain comets coincide exactly with those of meteor streams as independently observed. Thus, Tempel's comet, seen in 1866, coincides with the November meteors, or Leonids; Biela's comet, with the Andromeda meteors; while the bright comet of 1862 coincides with the August Perseids. Seventy such cases of the association of comets and meteor streams are now known; and Professor Lockyer has completed the proof of the connection by showing, that, when fragments of meteoric stones are intensely heated in a vacuum they afford a spectrum closely resembling those of comets. Some meteors are visible every fine night, and it has been calculated by Professor Newton of Yale College that seven and a half millions enter the earth's atmosphere every day; and if we add to these the much greater number that must escape observation, it is supposed that the actual number may be several hundred millions. Of course it is only by a kind of accident that the orbit of our earth crosses any of these meteoric streams, so that there are certainly a vast number, perhaps thousands or even millions, of such streams in the solar system, since some hundreds are either known or suspected to cross our path. Taking into consideration these numerous meteor-streams moving in elliptic orbits round the sun, together with the vast number of stray meteors, as it were, indicated by those that are seen every day in the year and by the exceptionally large and rare fireballs, we are led to the conclusion that the space occupied by the solar system, instead of being almost empty, as formerly supposed, is really full of solid bodies varying in size from that of dust or sand-grains up to huge masses a thousand times that of our earth.

The eight major planets are so remote from each other that if we represent the solar system as an open plain two and a half miles in diameter, our earth will in due proportion be shown by a pea, Mars by a large pin's head, Jupiter by an orange, and Neptune on the extreme outer edge by a largish plum. From any one of them the nearest would be invisible to us unless brilliantly illuminated, and however smooth and open was the plain, we might walk across it again and again in every direction, and with the exception of the two-foot ball in the centre representing the sun, we should probably declare it to be absolutely empty. Looking thus at the solar system, the vast emptiness, the absurd disproportion of the sizes of the planets to the immense spaces around and between them, was almost oppressive; and even when we took account of the nebular hypothesis, and tried to imagine a mass of elemental gaseous matter occupying a sphere of the diameter of the orbit of Neptune, gradually cooling and shrinking, leaving rings of diffused matter behind it, which afterwards broke up and aggregated into the planets and satellites already known to us, the hypothetical solution of the problem was hardly satisfying, since it seemed difficult to understand how so vast a *plenum* could be converted into an equally vast *vacuum*, except for the few and remotely scattered planetary systems as its sole relics.

But the study of the long-despised and misunderstood meteorites and falling stars has entirely changed our conceptions of that portion of the universe of which our sun is the centre. We are now led to regard it as more nearly approaching a plenum than a vacuum. We know that it is everywhere full of what may be termed planetary and meteoric life—full of solid moving bodies forming systems of various sizes and complexities from the vast mass of Jupiter with its five moons, down to some of the minor planets a few miles in diameter, and just large enough to become visible by reflected light; and again, downward, of all lesser dimensions to the mere dust-grains which only become visible when the friction on entering our atmosphere with the great velocities due to their planetary motion round the sun ignites and sometimes, perhaps, dissipates them.

We here obtain a new conception of the possible origin of the universe as we now see it (a conception which originated with Professor Tait, and has been forcibly advocated by Lockyer and a few other astronomers), which is, that both the solar system and the stellar universe have arisen from the aggregation of widely diffused solid particles, molecules, or atoms, whose coming together under the influence of gravitation produces heat, incandescence, and sometimes elemental vaporization, rather than from a primitive cosmic vapour from which solid masses have been formed by cooling and contraction. Everywhere we become aware of these solid masses of various sizes occupying the spaces around us. The rings of Saturn are composed of such solid particles in a state of unusual condensation. The vast ring of the minor planets indicates probably the existence there of millions of smaller invisible bodies forming a stream of meteors, analogous to some of those which cross our orbit but which are composed of much smaller bodies since none of them are independently visible. Then we have the comets, consisting of a dense swarm of such meteors whose frequent collisions may produce the luminous gases indicated by their spectra. Yet further, the strange zodiacal light, extending from the sun beyond the earth's orbit, is well explained as due to the light reflected under favourable conditions from the countless streams of meteors ever increasing in density as they approach the sun.

In its wider application to the stellar universe, the same theory serves to explain phenomena once supposed to be radically distinct. There is now known to be a perfect gradation from the faintest and least condensed nebulæ to the most brilliant stars, and these are all explained, on what is termed the meteoritic hypothesis, as being different stages in the aggregation of meteoritic matter everywhere and always going on. From the faintest diffused nebulæ we pass to those which exhibit a radial or spiral mode of condensation, and to others which possess a dense nucleus like a comet; then we have the compact discs called planetary nebulæ, and others which seem to be aggregated around one or more bright stars. Recently it has been found that many stars,—among others those of the Pleiades—which appear as stars only in the most powerful telescopes, are really nebulous stars when photographed with very long exposure under conditions which exhibit many thousands of stars which no telescope can render visible. And when these various bodies are examined with the spectroscope, they are seen to have many features in common, such as indicate differences in temperature and consequent difference in the amount and character of the luminous gases due to their greater or less condensation. The nebulæ of various forms and intensity represent, therefore, the early stages in the development of stars, suns, and planetary systems out of diffused meteoritic matter; while stars themselves are of various temperatures, the heat increasing when the meteoritic matter is most rapidly aggregating, and afterwards cooling till they become of so low a temperature as to cease to be luminous to our vision, as is the case with the dark companions of some of the spectroscopic double stars.

This conception of the meteoritic constitution of the whole stellar universe is one of the grandest achievements of the science of the nineteenth century. All the other astronomical discoveries of the period (except those gained through the spectroscope) are additions to our knowledge of essentially the same nature as others which preceded them; but in this case we have a new and comprehensive generalisation which links together a vast host of phenomena which, till quite recently, were isolated or misunderstood.

Beginning with the meteoric masses which at considerable intervals fall upon the earth, and the meteoric or cosmic dust which in minute spherules is probably always falling—since it is found abundantly in all the deepest oceanic deposits far removed from continental land—we have next the meteor streams with their attendant comets, circling round the sun in vast numbers, and increasing to such an extent in his vicinity that they become visible as the zodiacal light; the planetoids, ever increasing in recorded numbers and probably forming the larger members of a vast meteor ring; and the rings of Saturn, now proved to be of the same meteoritic nature. Then, passing on to the interstellar spaces, we find the nebulæ, which are but vast uncondensed meteor swarms; the planetary nebulæ and nebulous stars being examples of greater condensation, leading on to the myriads of the starry hosts, each one a sun heated by the inward rush and titanic collisions of countless meteor-swarms. These suns, after reaching a maximum of heat and light, slowly cool into darkness, until a collision with other cosmic matter again heats the mass to incandescence or even to vaporization—all this grand series of phenomena, rising from dust particles on the ocean bed to a million million of suns, comprehended, and to some extent explained, by one of the simplest and at first sight most inadequate of hypotheses—that of a meteoritic origin of the material universe.

It has been objected that this theory is not so simple as the old nebular hypothesis, and has no advantages over it. But this is a mistake. The latter begins with what we now see to be an impossible condition—that of a universe in a state of vapour. For all matter, in the absence of heat, is solid; and the only sources of heat we know of are, impact or friction, and chemical combination including electric action. Heat, therefore, in all its degrees and manifestations, will necessarily arise from diffused solid matter subject to gravitation, but it will arise partially and locally, not universally; and we now know that there *are* such varieties of temperature in the stellar universe. We have also positive evidence of solid matter everywhere, in an almost infinite gradation of size and of temperature, from that amount of cold in which no liquid, and perhaps no gas, can exist, up to that amount of heat in which all the elements are vaporized. We *can* conceive how, from diffused solid matter, without heat, the actual condition of the universe may have arisen; but we *cannot* conceive any previous condition which would result in the universal vaporization of all matter which the nebular hypothesis presupposes.

But this grand meteoritic theory, like all possible theories or speculations as to the origin of the cosmos, only takes us one step backward, and then leaves us no whit nearer to a real comprehension of the great insoluble problem. For we ask whence came this inconceivably vast extension of meteoric matter? What was its antecedent state? How did matter, at first presumably simple or atomic, aggregate into those forms we know as elements? And even if we could get back to a universe of primitive atoms, we should still be no nearer a complete solution, for then would begin a new series of questions far more difficult to answer. We should begin to seek after the origin of the FORCES which caused the development of atoms into matter and into worlds. Whence the simplest cohesive forces? Whence the chemical forces? And more mysterious than all, whence the force of gravitation, infinite, unchangeable, and at the very root of cosmic development? Beyond these problems again, and quite as essential and insoluble, are the problems of the ether. What is the ether, and what are its relations to matter? Whence the forces that cause the ether to vibrate, and in the various forms of heat, light, or electricity to be the source of all change of form, all molecular motion, all those infinite modifications in the states of matter that alone seem to render possible the development of organized living forms? To all these questions we have no definite answers, and probably never shall have; but we have at least one suggestive speculation, that of the vortex-theory of matter.

According to this theory, the ether is an incompressible frictionless fluid, and is the one and only substance of the universe. Matter is but a form of motion of the ether. Atoms are minute vortices, or rapidly revolving portions of the ether, which, when once started in this frictionless fluid, are eternal and indestructible. A sufficient number (almost infinite) of these vortices, of various dimensions and spinning with various velocities, and having progressive motions in every possible direction like the molecules of a gas, will, it is suggested, group themselves into various aggregations according to similarities of size and motion, will thus produce the elements, which elements will act upon each other in the various modes of chemical combination, and thus will arise all the forms of molecular matter. But the continued motions of these atoms and their combinations will set up in the unmodified ether the special vibrations of heat and electricity, which, reacting on matter, will lead to that vast series of co-ordinated changes we recognise as the laws and phenomena of nature. Whether gravitation could possibly arise from the initial impulse given to the ether is doubtful; but in this vortex-theory, of which Lord Kelvin is the chief exponent in this country, we have the most important attempt yet made to get near to the beginnings of the universe. It is, of course, essentially inconceivable, as are all fundamental conceptions. The incompressible, frictionless, universal fluid is inconceivable; the origin of its infinity of atomic vortex motions is inconceivable; as are the translatory motions, the infinity of combinations, the complexity of chemical actions, the production of the varied kinds of ether-vibrations, and of gravitative force; and when we have fully grasped all these inconceivabilities there remains the still greater inconceivability of how life, consciousness, affection, intellect, arose from this infinite clash of ethereal vortex-rings!

The conception is, however, a grand one; and, together with the meteoritic hypothesis as to the immediate antecedents of the visible universe, must rank among the great intellectual achievements of our century. Yet they bring us no nearer to the First Cause of this vast cosmos in which we live; and most minds will feel that we never can get nearer to it than in "the consciousness of an Inscrutable Power manifested to us through all phenomena," which Herbert Spencer considers to be the logical and the utmost outcome of the most far-reaching human Science.

## CHAPTER XII

### GEOLOGY; THE GLACIAL EPOCH, AND THE ANTIQUITY OF MAN

The hills are shadows, and they flow
  From form to form, and nothing stands;
  They melt like mist, the solid lands,
Like clouds they shape themselves and go.

—*Tennyson.*

With cunning hand he shapes the flint,
  He carves the bone with strange device,
He splits the rebel rock by dint
  Of effort—till one day there flies
A spark of fire from out the stone,
Fire, which shall make the world his own.

—*Mathilde Blind.*

THE foundations of modern geology were laid, in the latter part of the last century, by Werner, Hutton and William Smith, but most of the details and some of the more important principles have been wholly worked out during the present century. The great landmarks of its progress can alone be referred to here, namely (1) the establishment by Lyell of what has been termed the uniformitarian theory; (2) the proof of a recent glacial epoch and the working out of its effects upon the earth's surface; and (3) the discovery that man in the northern hemisphere lived contemporaneously with many now extinct animals.

In the early part of the century, and so late as the year 1830, Cuvier's *Essay on the Theory of the Earth* held the field as the exponent of geological theory. A fifth edition of the English translation appeared in 1827, and a German translation so late as 1830. In this work it was maintained that almost all geological phenomena pointed to a state of the earth and of natural forces very different from what now exists. In the raised beds of shells, in fractured rocks, in vertical stratification, we were said to have proofs "that the surface of the globe has been broken up by revolutions and catastrophes." The differences in the character of adjacent stratified deposits showed that there must have been various successive irruptions of the sea over the land; and Cuvier maintained that these irruptions and retreats of the sea were not slow or gradual, "but that most of the catastrophes which have occasioned them have been sudden." He urged that the sharp and bristling ridges and peaks of the primitive mountains "are indications of the violent manner in which they have been elevated;" and he concludes that "it is in vain we search among the powers which now act at the surface of the earth for causes sufficient to produce the revolutions and catastrophes, the traces of which are exhibited in its crust." This theory of convulsions and catastrophes held almost universal sway within the memory of persons now living, for although Hutton and Playfair had advanced far more accurate views, they appear to have made little impression, while the great authority attached to Cuvier's name carried all before it.

But in 1830, while Cuvier was at the height of his fame, and his book was still being translated into foreign languages, a hitherto unknown writer published the first volume of a work which struck at the very root of the catastrophic theory, and demonstrated by a vast array of facts and the most cogent reasoning, that almost every portion of it was more or less imaginary and in opposition to the plainest teachings of nature. The victory was complete. From the date of the publication of the *Principles of Geology* there were no more English editions of *The Theory of the Earth.*

Lyell's method was that of a constant appeal to the processes of nature. Before asserting that certain results could not be due to existing causes he carefully observed what those causes were now doing. He applied to them the tests of accurate measurement, and he showed that, taking into account the element of long-continued action, they were, in almost every case, fully adequate to explain the observed phenomena. He showed that modern volcanoes had poured out equally vast masses of melted rock, which had covered equally large areas, with any ancient volcano; that strata were now forming, comparable in extent and thickness with any ancient strata; that organic remains were being preserved in them, just as in the older formations; that land was almost everywhere either rising or sinking, as of old; that valleys were being excavated and mountains worn away; that earthquake shocks were producing faults in the rocks; that vegetation was now preparing future coal-beds; that limestones, sandstones, metamorphic and igneous rocks were still being formed; and that, given time, and the intermittent or continuous action of the causes we can now trace in operation, all the contortions and fractures of strata, all the ravines and precipices, and every other modification of the earth's crust supposed to imply the agency of sudden revolutions and violent catastrophes may be again and again produced.

During a period of more than forty years Sir Charles Lyell continued to enlarge and improve his work, bringing out eleven editions, the last of which was published three years before his death; and rarely has any scientific work so completely justified its title, since it remains to this day the best exposition of the *Principles of Geology*—the foundation on which the science itself must be and has been built. The disciples and followers of Lyell have been termed "Uniformitarians," on account of their belief that the causes which produced the phenomena manifested to us in the crust of the earth are essentially of the same nature as those acting now. And, as is often the case, the use of the term as a nickname has led to a misconception as to the views of those to whom it is applied. A few words on this point are therefore called for.

Modern objectors say, that it is unphilosophical to maintain that in our little experience of a few hundred, or at most a few thousand, years, we can have witnessed all forms and degrees of the action of natural forces; that we have no right to take the historical period as a fair sample of all past geological ages; and that, as a mere matter of probability, we ought to expect to find proofs of greater earthquakes, more violent eruptions, more sudden upheavals, and more destructive floods, having occurred during the vast eons of past time. Now this argument is perfectly sound if limited to the occurrence of extreme cases, but not if applied to averages. No uniformitarian will deny the probability of there having been *some* greater convulsions in past geological ages than have ever been experienced during the historical period. But modern convulsionists do not confine themselves to this alone, but maintain that, *as a rule*, all the great natural forces tending to modify the surface of the earth were more powerful and acted on a larger scale than they do now. On the ground of mere probability, however, we have no right to assume a diminution rather than an increase of natural forces in recent times, unless there is some proof that these forces have diminished. Sir Charles Lyell shows that the cases adduced as indicating greater forces in the past are fallacious, and his doctrine is simply one of real as against imaginary forces.

But our modern objectors have another argument, founded upon the admitted fact that the earth has cooled and is slowly cooling, and was probably once in a molten condition. They urge that in early geological times, when the earth was hotter, the igneous, aqueous, and aerial forces, were necessarily greater, and would produce more rapid changes and greater convulsions than now. This is a purely theoretical conclusion, by no means sure, and perhaps the very reverse of what really occurred. There are two reasons for this belief, which may be very briefly stated. After the earth's crust was once formed it cooled very slowly, and the crust became very gradually thicker. So far as the action of the molten interior on the crust may have produced convulsions they should become not less, but more violent as the crust becomes thicker. With a thin crust any internal tension will be more frequently relieved by fracture or bending, and the resulting disturbances will be *less* violent; but as the crust becomes thicker, internal tensions will accumulate, and when relieved by fracture the disturbance will be *more* violent.

As regards storms and other aerial disturbances, these also would probably be less violent when the temperature of the whole surface was more uniform as well as warmer, and the atmosphere consequently so full of vapour as to prevent the sun's rays from producing the great inequalities of temperature that now prevail. It is these inequalities that produce the great aerial disturbances of our era, which arise from the heated surfaces of the bare plains and deserts of the sub-tropical and warm-temperate belts. In the equatorial belt (10° each side of the equator), where the heat is more uniform and the surface generally well clothed with vegetation, tornadoes and hurricanes are almost unknown.

There remains only the action of the tides upon coasts and estuaries, which may have been greater in early geological times, if, as is supposed, the moon was then considerably nearer to the earth than it is now. But this is a comparatively unimportant matter as regards geological convulsions, because its maximum effects recur at short intervals and with great regularity, so that both vegetation and the higher forms of animal life would necessarily be limited to the areas which were beyond its influence.

It thus appears that, so far from there being any theoretical necessity for greater violence of natural forces in early geological times, there are some weighty reasons why the opposite should have been the case; while all the evidence furnished by the rocks themselves, and by the contours of the earth's surface, are in favour of a general uniformity, with, of course, considerable local variability.

It is interesting to note the very different explanations of the commonest features of the earth's surface given by the old and by the new theories. In every mountain region of the globe deep valleys, narrow ravines and lofty precipices are of common occurrence, and these were, by the old school, almost always explained as being due to convulsions of nature. In ravines, we were taught that the rocks had been "torn asunder," while the mountains and the precipices were indications of "sudden fractures and upheavals of the earth's crust." On the new theory, these phenomena are found to be almost wholly due to the slow action of the most familiar every-day causes, such as rain, snow, frost, and wind, with rivers, streams, and every form of running water, acting upon rocks of varying hardness, permeability and solubility. Every shower of rain falling upon steep hill-sides or gentle slopes, while partially absorbed, to a large extent runs over the surface, carrying solid matter from higher to lower levels. Every muddy stream or flooded river shows the effect of this action. Day and night, month after month, year after year, this denudation goes on, and its cumulative effects are enormous. The material is supplied from the solid rocks, fractured and decomposed by the agency of snow and frost or by mere variations of temperature, and primarily by those interior earth-movements which are continually cleaving, fissuring and faulting the solid strata, and thus giving the superficial causes of denudation facilities for action. The amount and rate of this superficial erosion and denudation of the earth's surface can be determined by the quantity of solid matter carried down by the rivers to the sea. This has been measured with considerable accuracy for several important rivers; and by comparing the quantity of matter, both in suspension and solution, with the area of the river basin, we know exactly the average amount of lowering of the whole surface per annum. It has thus been calculated that—

| The Mississippi removes one foot of the surface of its basin in . . . . . . | 6,000 years. |
|---|---|
| „ Ganges „ „ „ „ | 2,358 „ |
| „ Hoang Ho „ „ „ „ | 1,464 „ |
| „ Rhone „ „ „ „ | 1,528 „ |
| „ Danube „ „ „ „ | 6,846 „ |
| „ Po „ „ „ „ | 729 „ |
| „ Nith „ „ „ „ | 4,723 „ |

The average of these rivers gives us one foot as the lowering of the land by sub-aerial denudation in 3,000 years, or a thousand feet in three million years; but as Europe has a mean altitude of less than a thousand feet, it follows that at the present rate of denudation the whole of Europe would be reduced to nearly the sea-level in about three million years. Before this method of measuring the rate of the lowering of continents was hit upon by Mr. Alfred Tylor in 1853, no one imagined that it was anything like so rapid; and, as a million years is certainly a short period as compared with the whole geological record, it is clear that elevation must, on the whole, have always kept pace with the two lowering agencies—sinking and denudation. Again, as in every continent the areas occupied by plains and lowlands, where denudation is comparatively slow, are large as compared with the mountain areas, where all the denuding agencies are most powerful, it is probable that most mountain ranges are being lowered at perhaps ten times the above average rate, and many mountain peaks and ridges perhaps a hundred times.

Examples of the rapidity of denudation as compared with earth-movements are to be found everywhere. In disturbed regions, faults of many hundreds, and sometimes even thousands of feet, are not uncommon; yet there is often no inequality on the surface, indicating that the dislocation of strata has been caused by small and often-repeated movements, at such intervals that denudation has been able to remove the elevated portion as it arose. Again, when the strata are bent into great folds or undulations, it is only rarely that the tops of the folds correspond to ridges and the depressions to valleys. Frequently the reverse is the case, a valley running along the anticlinal line or structural ridge, while the synclinal or structural hollow forms a mountain top; while in other cases, valleys cut across these structural features, with little or no regard to them. This results from the fact that it is not mountains or mountain ranges, as we see them, which have been raised by internal forces, but a considerable area, already perhaps much disturbed and dislocated by earth-movements, has been slowly raised till it became a kind of table-land. From its first elevation above the sea, however, it would have been exposed to rainfall, and the water flowing off in the direction of least resistance would have formed a number of channels radiating from the highest portion, and thus establishing the first outlines of a system of valleys, which go on deepening as the land goes on rising, often quite irrespective of the nature of the rocks beneath. This explains the close resemblance in the general arrangement of valleys in all high regions, as well as the very common phenomenon of a river crossing the main range of a mountain system by a deep gorge; for this merely shows that what is now the highest part of the range was at first lower than that where the river has its source, but has become higher by the more rapid degradation of the lateral ranges, owing to their being formed of rock which is more easily disintegrated. The various peculiarities of open valley and narrow gorge, of sloping mountain side or lofty precipice, of rivers cutting across hills, as in the South Downs and at Clifton, when open plains by which they might apparently have reached the sea are near at hand, may be all explained as the results of those simple causes which are everywhere in action around us. It was Sir Charles Lyell who first convinced the whole scientific world of the efficacy of these familiar agents; and the secure establishment of this doctrine constitutes one of the great philosophical landmarks of the Nineteenth Century.

### *The Glacial Epoch.*

The proof of the recent occurrence in the north-temperate zone of a glacial epoch, during which large portions of Europe and North America were buried in ice, may, from one point of view, be thought to prove that other agents than those now in operation have acted in past ages, and thus to disprove the main assumption of the uniformitarians. But, on the other hand, its existence has been demonstrated by those very methods which Sir Charles Lyell advocated—the accurate observation of what nature is doing now; while an ice age really exists at the present time in Greenland, in the same latitude as nearly the whole of

Sweden and Norway, which enjoy a comparatively mild climate.

The first clear statement of the evidence for a former ice age was given, in 1822, by a Swiss engineer named Venetz. He pointed out that, where the existing glaciers have retreated, the rocks which they had covered are often rounded, smoothed, and polished, or grooved and striated in the direction of the glacier's motion; and that, far away from any existing glaciers, there were to be seen rocks similarly rounded, polished and striated; while there also existed old moraine heaps exactly similar to those formed at present; and that these phenomena extended as far as the Jura range, on the flanks of which there were numbers of huge blocks of stone, of a kind not found in those mountains but exactly similar to the ancient rocks of the main Alpine chain. Hence, he concluded that glaciers formerly extended down the Rhone valley as far as the Jura, and there deposited those erratic blocks, the presence of which had puzzled all former observers.

Soon afterwards, Charpentier and Agassiz devoted themselves to the study of the records left by the ancient glaciers; and from that time to the present a band of energetic workers in every part of the world have, by minute observation and reasoning, established the fact of the extension of glaciers, or ice-sheets, over a large portion of the north-temperate zone; and have also determined the direction of their motion and the thickness of the ice in various parts of their course. These conclusions are now admitted by every geologist who has devoted himself to the subject, and are embodied in the various official geological surveys of the chief civilized countries; and as they constitute one of the most remarkable chapters in the past history of the globe, and especially as this great change of climate occurred during the period of man's existence on the earth, a brief sketch of the facts must be here given.

There are four main groups of phenomena which demonstrate the former existence of glaciers in areas where they now now absent: (1) Moraines, and glacial drifts or gravels; (2) Smoothed, rounded or planed rocks; (3) Striæ, grooves, and furrows on rock-surfaces; (4) Erratics and perched blocks.

(1) Moraines are formed by all existing glaciers, consisting of the earth and rocks which fall upon the ice-rivers from the sides of the valleys through which they flow. The slow motion of the glacier carries these down with it, and they are deposited in great heaps where it melts. In some glaciers, where the tributary valleys are numerous and the *débris* that falls upon the ice is abundant, the whole of the lower part of the glacier for many miles is so buried in it that the surface of the ice cannot be seen, and in these cases there will be a continuous moraine formed across the valley where the glacier terminates. The characteristics of moraines are, that they consist of varied materials, earth, gravel, and rocks of various sizes intermingled confusedly; and they often form mounds or ridges completely across a valley, except where the stream passes through it, while in other cases they extend laterally along the slopes of the hill-sides, where, owing to the form of the valley, the glacier has shrunk laterally and left its lateral moraine behind it. In many cases huge blocks of rock rest on the very summit of a moraine, or, in the case of lateral moraines, on the very edge of a precipice in positions where no known agency but

ice could have deposited them. These are called "perched blocks." Drifts or glacial gravels are deposits of material similar to that forming the moraines, but spread widely over districts which have formerly been buried in ice. These are often partially formed of stiff clay, in which are imbedded quantities of smoothed and striated stones; but the great characteristic of all these ice-products is that the materials are not stratified,—that is, sorted according to their fineness or coarseness, as is always the case when deposited by water,—but are mingled confusedly together, the large stones being scattered all through the mass, and usually being quite as abundant at the top as at the bottom of the deposit. Such deposits are to be found all over the north and north-west of our islands, and are often well exhibited in railway cuttings; and wherever they are well developed, and the materials of which they consist differ from those forming the underlying rocks, they are an almost infallible indication of the former existence of a glacier or ice-sheet.

(2) The smoothed and rounded rocks, called in Switzerland *roches moutonnées*, from their resemblance at a distance to recumbent sheep, are present in almost all recently-glaciated mountainous countries, especially where the rocks are very hard. They are to be seen in all the higher valleys of Wales, the Lake District, and Scotland, and on examination are found to consist often of the hardest and toughest rocks. In other cases the rock forming the bed of the valley is found to be planed off smooth, even when it consists of hard crystalline strata thrown up at a high angle, and which naturally weathers into a jagged or ridged surface.

(3) The smoothed rocks are often found to be covered with numerous striæ, deep grooves, or huge flutings, and these are almost always in one direction, which is that of the course of the glacier. They may often be traced in the same direction for miles, and do not change in harmony with the lesser inequalities of the valley, as they would certainly do had they been formed by water action. These striæ and smoothed rocks are often found hundreds or even thousands of feet above the floor of the valley, and in many cases a definite line can be traced, above which the rocks are rugged and jagged, while below it they are all more or less rounded, smooth, or polished.

(4) Erratic blocks are among the most widespread and remarkable indications of glacial action, and they were the first that attracted the attention of men of science. The great plains of Denmark, Prussia, North Germany, and Russia are strewn with large masses of granite and hard metamorphic rocks, and these rest either on glacial drift or on quite different rocks of Secondary or Tertiary age. In parts of North Germany they are so abundant as to hide the natural surface, and they are often piled up in irregular heaps forming hills of granite boulders covered with forests of pine, birch, and juniper. Many of these blocks are more than a thousand tons weight, and almost all of them can be traced to the mountains of Scandinavia as their source. Many of the largest blocks have been carried furthest from the parent rock—a fact which is conclusive against their having been brought to their present position by the action of floods.

The most interesting and instructive erratic blocks

are those found upon the slopes of the Jura, because they have been most carefully studied by Swiss and French geologists, and have all been traced to their sources in the Alpine chain. The Jura mountains consist wholly of Secondary limestones, and are situated opposite to the Bernese Alps, at a distance of about fifty miles. Along their slopes for a distance of a hundred miles, and extending from their base to a height of 2,000 feet above the Lake of Neuchâtel, are great numbers of rocks, some of them as large as houses, and always quite different from that of which the Jura range is formed. These have all been traced to their parent rocks in various parts of the course of the old glacier of the Rhone, and, what is even more remarkable, their distribution is such as to prove that they were conveyed by a glacier and not by floating ice during a period of submergence. The rocks and other *débris* that fall upon a glacier from the two sides of its main valley form distinct moraines upon its surface, and however far the glacier may flow and however much it may spread out where the valley widens, they preserve their relative position so that whenever they are deposited by the melting of the glacier those that came from the north side of the valley will remain completely separated from those which came from the south side. It was this fact which convinced Sir Charles Lyell that the theory of floating ice, which he had first adopted, would not explain the distribution of the erratics, and he has given in his *Antiquity of Man* (4th ed., p. 344) a map showing the course of the blocks as they were conveyed on the surface of the glacier to their several destinations. Other blocks are found on the lower slopes of the Alpine chain towards Bern on one side and Geneva on the other, while the French geologists have traced them down the Rhone valley seventy miles from Geneva, and also more than twenty miles west of the Jura, thus proving that at the lowest portion of that chain the glacier flowed completely over it. In all these cases the blocks can be traced to a source corresponding to their position on the theory of glacier action. Some of these rocks have been carried considerably more than 200 miles, proving that the old glacier of the Rhone extended to this enormous distance from its source.

In our own islands and in North America these various classes of evidence have been carefully studied, the direction of the glacial striæ everywhere ascertained, and all the more remarkable erratic blocks traced to their sources, with the result that the extent and thickness of the various glaciers and ice-sheets are well determined and the direction of motion of the ice ascertained. The conclusions arrived at are very extraordinary, and must be briefly indicated.

In Great Britain, during the earlier and later phases of the ice age, all the mountains of Scotland, the Lake District, and Wales produced their own glaciers, which flowed down to the sea. But at the time of the culmination of the Glacial Epoch the Scandinavian ice-sheet extended on the south-east till it filled up the Baltic Sea and spread over the plains of north-western Europe, and also filled up the North Sea, joining the glaciers of Scotland, forming with them a continuous ice-sheet from which the highest mountains alone protruded. At the same time this Scotch ice-sheet extended into the Irish Sea, and united with the glaciers of the Lake District, Wales, and Ireland till almost continuous ice-sheets enveloped those countries also. Glacial striæ are found up to a height of 3,500 feet in Scotland and 2,500 feet in the Lake District and in Ireland; while the Isle of Man was completely overflowed, as shown by glacial striæ on the summit of its loftiest mountains. Erratics from Scandinavia are found in great quantities on Flamborough Head, mixed with others from the Lake District and Galloway, showing that two ice-streams met here from opposite directions. Erratics from Scotland are also found in the Lake District, in North Wales, in the Isle of Man, and in Ireland, from which the direction of the moving ice can be determined. Great numbers of local rocks have also been carried into places far from their origin, and in every case this displacement is in the direction of the flow of the ice as ascertained by the other evidence—never in the opposite direction. Each great mountain area had, however, its own centre of local dispersal, depending upon the position of greatest thickness of the ice-sheet, which was not necessarily that of the highest mountains, but was approximately the centre of the main area of glaciation. Thus the centre of the North Wales ice-sheet was not at Snowdon, but over the Arenig mountains, which thus became a local centre of dispersal of erratics. In Ireland, the mountains being placed around the coasts, the great central plain became filled with ice which, continually accumulating, formed a huge dome of ice whose outward pressure caused motion in all directions till checked by the opposing motion of the great Scandinavian ice-sheet. This strange fact has been demonstrated by the work of the Irish Geological Survey and by many local geologists, and is universally accepted by all who have studied the evidence.

The great outlines of the phenomena of the ice age in our islands are now as thoroughly well established as any of the admitted conclusions of geological science. In our own country the ice extended more or less completely over the whole of the midland counties and as far south as the Thames Valley.

When we cross the Atlantic the phenomena are equally remarkable. The whole of the north-eastern United States and Canada were also buried in an ice-sheet of enormous thickness and extent. It came southward as far as New York, and inland, in an irregular line, by Cincinnati, to St. Louis on the Mississippi. The whole of the region to the north of this line is covered with a deposit of drift, often of enormous thickness, while, embedded in the drift or scattered over its surface, are numbers of blocks and rock-masses, often formed of materials quite foreign to the bed-rock of the district. These erratics have in many cases been traced to their sources, sometimes 600 miles away, and the study of these, and of the numerous grooved and striated rocks, show that the centre of dispersal was far north of the Alleghanies and its outliers, and, as in the case of Ireland, must have consisted of a huge dome of ice situated over the plateau to the north of the great lakes, in what must have been an area of great snow-fall combined with a very low temperature. The maximum thickness of this great ice-sheet must have been at least a mile over a considerable portion of its area, as glacial deposits have been found on the summit of Mount Washington at an altitude of nearly 6,000 feet, and the centre of motion was a considerable distance to the north-west, where it must have reached a still greater altitude.

The complete general similarity of the conclusions reached by four different sets of observers in four different areas—Switzerland, North-western Europe, the British Isles, and North America—after fifty years of continuous research, and after every other less startling theory had been put forth and rejected as wholly inconsistent with the phenomena to be explained, renders it as certain as any conclusion from indirect evidence can be, that a large portion of the North temperate zone, now enjoying a favourable climate and occupied by the most civilized nations of the world, was, at a very recent epoch, geologically speaking, completely buried in ice just as Greenland is now. How recently the ice has passed away is shown by the perfect preservation of innumerable moraines, perched blocks, erratics, and glaciated rock-surfaces, showing that but little denudation has occurred to modify the surface; while undoubted relics of man found in glacial or interglacial deposits prove that it occurred during the human period. It is clear that man could not have lived in any area while it was actually covered by the ice-sheet, while any indications of his presence at an earlier period would almost certainly be destroyed by the enormous abrading and grinding power of the ice.

Besides the areas above referred to, there are wide-spread indications of glaciation in parts of the world where a temperate climate now prevails. In the Pyrenees, Caucasus, Lebanon, and Himalayas glacial moraines are found far below the lower limits they now attain. In the Southern Hemisphere similar indications are found in New Zealand, Tasmania, and the southern portion of the Andes; but whether this cold period was coincident with that of the Northern Hemisphere we have at present no means of determining, nor even whether they were coincident among themselves, since it is quite conceivable that they may have been due to local causes, such as greater elevation of the land, and not to any general cause acting throughout the South temperate zone.

In the North temperate zone, however, the phenomena are so widespread and so similar in character, with only such modifications as are readily explained by proximity to, or remoteness from, the ocean, that we are almost sure they must have been simultaneous, and have been due to the same general causes, though perhaps modified by local changes in altitude and consequent modification of winds or ocean-currents. The time that has elapsed since the glaciation of the Northern Hemisphere passed away is, geologically, very small indeed, and has been variously estimated at from 20,000 to 100,000 years. At present the smaller period is most favoured by geologists, but the duration of the ice-age itself, including probably one or more inter-glacial mild periods, is admitted to be much longer, and probably to approach the higher figure above given.

The undoubted fact, however, that a large part of the North Temperate zone has been recently subjected to so marvellous a change of climate, is of immense interest from many points of view. It teaches us in an impressive way how delicate is the balance of forces which renders what are now the most densely peopled areas habitable by man. We can hardly suppose that even the tremendously severe ice age of which we have evidence is the utmost that can possibly occur; and, on the other hand, we may anticipate that the condition of things which in earlier geological times rendered even the polar regions adapted for a luxuriant woody vegetation, may again recur, and thus vastly extend the area of our globe which is adapted to support human life in abundance and comfort. In the endeavour to account for the change of climate and of physical geography which brought about so vast a change, and then, after a period certainly approaching, and perhaps greatly exceeding, a hundred thousand years, caused it to pass away, some of the most acute and powerful intellects of our day have exerted their ingenuity; but, so far as obtaining general acceptance for the views of any one of them, altogether in vain. There seems reason to believe, however, that the problem is not an insoluble one; and when the true cause is reached, it will probably carry with it the long-sought datum from which to calculate with some rough degree of accuracy the duration of geological periods. But, whether we can solve the problem of its cause or no, the demonstration of the recent occurrence of a Glacial Epoch or Great Ice Age, with the determination of its main features over the Northern Hemisphere, will ever rank as one of the great scientific achievements of the Nineteenth Century.

### *The Antiquity of Man.*

Following the general acceptance of a glacial epoch by about twenty years, but to some extent connected with it, came the recognition that man had existed in Northern Europe along with numerous animals which no longer live there—the mammoth, the woolly rhinoceros, the wild horse, the cave-bear, the lion, the sabre-toothed tiger, and many others—and that he had left behind him, in an abundance of rude flint implements, the record of his presence. Before that time geologists, as well as the whole educated world, had accepted the dogma that man only appeared upon the earth when both its physical features and its animal and vegetable forms were exactly as we find them to-day; and this belief, resting solely on negative evidence, was so strongly and irrationally maintained that the earlier discoveries could not get a hearing. A careful but enthusiastic French observer, M. Boucher de Perthes, had for many years collected with his own hands, from the great deposits of old river gravels in the valley of the Somme near Amiens, abundance of large and well-formed flint implements. In 1847 he published an account of them, but nobody believed his statements, till, ten years later, Dr. Falconer, and, shortly afterwards, Professor Prestwich and Mr. John Evans, examined the collections and the places where they were found, and were at once convinced of their importance; and their testimony led to the general acceptance of the doctrine of the great antiquity of the human race. From that time researches on this subject have been carried on by many earnest students, and have opened up a number of altogether new chapters in human history.

So soon as the main facts were established, many old records of similar discoveries were called to mind, all of which had been ignored or explained away on account of the strong prepossession in favour of the very recent origin of man. In 1715 flint weapons had been found in excavations near Gray's Inn Lane, along with the skeleton of an elephant. In 1800 another discovery was made in Suffolk of flint weapons and the remains of extinct animals in the same deposits.

In 1825 Mr. McEnery, of Torquay, discovered worked flints along with the bones and teeth of extinct animals in Kent's cavern. In 1840 a good geologist confirmed these discoveries, and sent an account of them to the Geological Society of London, but the paper was rejected as being too improbable for publication! All these discoveries were laughed at or explained away, as the glacial striæ and grooves so beautifully exhibited in the Vale of Llanberris were at first endeavoured to be explained as the wheel-ruts caused by the chariots of the ancient Britons! These, combined with numerous other cases of the denial of facts on *a priori* grounds, have led me to the conclusion that, whenever the scientific men of any age disbelieve other men's careful observations without inquiry, the scientific men are *always* wrong.

Even after these evidences of man's great antiquity were admitted, strenuous efforts were made to minimise the time as measured by years; and it was maintained that man, although undoubtedly old, was entirely post-glacial. But evidence has been steadily accumulating of his existence at the time of the glacial epoch, and even before it; while two discoveries of recent date seem to carry back his age far into pre-glacial times. These are, first, the human cranium, bones, and works of art which have been found more than a hundred feet deep in the gold-bearing gravels of California, associated with abundant vegetable remains of extinct species, and overlaid by four successive lava streams from long extinct volcanoes. The other case is that of rude stone implements discovered by a geologist of the Indian Survey in Burma in deposits which are admitted to be of at least Pliocene Age. In both these cases the evidence is disputed by some geologists, who seem to think that there is something unscientific, or even wrong, in admitting evidence that would prove the Pliocene age of any other animal to be equally valid in the case of man. There is assumed to be a great improbability of his existence earlier than the very end of the Tertiary epoch. But all the indications drawn from his relations to the anthropoid apes point to an origin far back in Tertiary time. For each one of the great apes—the gorilla, the chimpanzee, the orang, and even the gibbons—resemble man in certain features more than do their allies, while in other points they are less like him. Now, if man has been developed from a lower animal form, we must seek his ancestors not in the direct line between him and any of the apes, but in a line towards a common ancestor to them all; and this common ancestor must certainly date back to the early part of the Tertiary epoch, because in the Miocene period anthropoid apes not very different from living forms have been found fossil.

There is therefore no improbability whatever in the existence of man in the later portions of the Tertiary period, and we have no right, scientifically, to treat any evidence for his existence in any other way than the evidence for the existence of other animal types.

It has been argued by some writers that, as no other living species of mammal goes back farther than the Newer Pliocene, therefore man is probably no older. But it is forgotten that the difference of man from the apes is not only specific but at least of generic or of family rank, while some naturalists place him even in a separate order of mammalia. Besides the erect posture and free hands, with all the details of anatomical structure which these peculiarities imply, the great development of his brain pre-eminently distinguishes him. We may suppose, therefore, that when he had reached the erect form and possessed all the external appearance of man, his brain still remained undeveloped, and the time occupied by this development was not improbably equal to that required for the specific modification of the lower mammalia. It is often forgotten that so soon as man used fire and made weapons, all further useful modification would be in the direction of increased brain power, by which he was able to succeed both in his struggle against the elements and with the lower animals. There is therefore no improbability in finding the remains or the implements of a low type of man in the early Pliocene period.

The certainty that man coexisted with many now extinct animals, and the probability of our discovering his remains in undoubted Tertiary strata, constitutes an immense advance on the knowledge and beliefs of our forefathers, and must therefore rank among the prominent features in the scientific progress of the Nineteenth Century.

# CHAPTER XIII

## EVOLUTION AND NATURAL SELECTION

Enkindled in the mystic dark,
 Life built herself a myriad forms,
And, flashing its electric spark
 Through films, and cells, and pulps, and worms,
Flew shuttlewise above, beneath,
Weaving the web of life and death.

—*Mathilde Blind.*

The world moves on in singing harmony—
 Her steps of eon length: from primal cloud,
 First through her realms old Chaos calls aloud;
Then, splashing in the Mesozoic sea,
Huge heralds of the beauty yet to be,
 Her saurian monsters rise; they pass away,
 And lo! the glories of a better day,
And man, the God-within, not fully free.

—*American Fabrian.*

We now approach the subject which, in popular estimation, and perhaps in real importance, may be held to be the great scientific work of the nineteenth century—the establishment of the general theory of evolution, by means of the special theory of the development of the organic world through the struggle for existence and its necessary outcome, Natural Selection. Although in the last century Buffon, Dr. Erasmus Darwin, and the poet Goethe, had put forth various hints and suggestions pointing to evolution in the organic world, which they undoubtedly believed to have occurred, no definite statement of the theory had appeared till early in the present century, when La Place explained his views as to the evolution of the stellar universe and of solar and planetary systems in his celebrated Nebular Hypothesis; and about the same time Lamarck published his *Philosophie Zoologique*, containing an elaborate exposition of his theory of the progressive development of animals and plants. But this theory gained few converts among naturalists, partly because Lamarck was before his time, and also because the causes he alleged did not seem adequate to produce the wonderful adaptations we everywhere see in nature. During the first half of the present century, owing to the fact that Brazil, South Africa, and Australia then became for the first time accessible to English collectors, the treasures of the whole world of nature were poured in upon us so rapidly, that the comparatively limited number of naturalists were fully occupied in describing the new species and endeavouring to discover true methods of classification. The need of any general theory of how species came into existence was hardly felt; and there was a general impression that the problem was at that time insoluble, and that we must spend at least another century in collecting, describing, and classifying, before we had any chance of dealing successfully with the origin of species. But the subject of evolution was ever present to the more philosophic thinkers, though the great majority of naturalists and men of science held firmly to the dogma that each species of animal and plant was a distinct creation, though how produced was admitted to be both totally unknown and almost, if not quite unimaginable.

The vague ideas of those who favoured evolution were first set forth in systematic form, with much literary skill and scientific knowledge, by the late Robert Chambers in 1844, in his anonymous volume, *Vestiges of the Natural History of Creation.* He passed in review the stellar and solar systems, adopted the Nebular Hypothesis, and sketched out the geological history of the earth, with continuous progression from lower to higher forms of life. After describing the peculiarities of the lower plants and animals, dwelling upon those features which seemed to point to a natural mode of production as opposed to an origin by special creation, the author set forth with much caution the doctrine of progressive development resulting from "an impulse which was imparted to the forms of life, advancing them in definite lines, by generation, through grades of organisation terminating in the highest plants and animals." The reasonableness of this view was urged through the rest of the work; and it was shown how much better it agreed with the various facts of nature and with the geographical distribution of animals and plants, than the idea of the special creation of each distinct species.

It will be seen, from this brief outline, that there was no attempt whatever to show *how* or *why* the various species of animals and plants acquired their peculiar characters, but merely an argument in favour of the reasonableness of the fact of progressive development, from one species to another, through the ordinary processes of generation. The book was what we should now call mild in the extreme. It was serious and even religious in tone, and calculated in this respect to disarm the opposition even of the most orthodox theologists; yet it was met with just the same storm of opposition and indignant abuse which assailed Darwin's work fifteen years later. As an illustration of the state of scientific opinion at this time, it may be mentioned that so great a man as Sir John Herschel, at a scientific meeting in London, spoke strongly against the book for its advocacy of so great a scientific heresy as the Theory of Development.

I well remember the excitement caused by the publication of the *Vestiges*, and the eagerness and delight with which I read it. Although I saw that it really offered no explanation of the process of change of species, yet the view that the change was effected, not through any unimaginable process, but through the known laws and processes of reproduction, commended itself to me as perfectly satisfactory, and as affording the first step towards a more complete and explanatory theory. It seems now a most amazing thing, that even to argue for this first step was accounted a heresy, and was almost universally condemned as being opposed to the teachings of both science and religion!

The book was, however, as great a success as, later on, was Darwin's *Origin of Species*. Four editions were issued in the first seven months, and by 1860 it had reached the eleventh edition, and about 24,000 copies had been sold. It is certain that this work did great service in familiarising the reading public with the idea of evolution, and thus preparing them for the more complete and efficient theory laid before them by Darwin.

During the fifteen years succeeding the publication of the *Vestiges* many naturalists expressed their belief in the progressive development of organic forms; while in 1852 Herbert Spencer published his essay contrasting the theories of Creation and Development with such skill and logical power as to carry conviction to the minds of all unprejudiced readers; but none of these writers suggested any definite theory of *how* the change of species actually occurred. That was first done in 1858; and in connection with it I may, perhaps, venture to give a few personal details.

Ever since I read the *Vestiges* I had been convinced that development took place by means of the ordinary process of reproduction; but though this was widely admitted, no one had set forth the various kinds of evidence that rendered it almost a certainty. I endeavoured to do this in an article written at Sarawak in February, 1855, which was published in the following September in the *Annals of Natural History*. Relying mainly on the well-known facts of geographical distribution and geological succession, I deduced from them the law, or generalisation, that, "Every species has come into existence coincident both in Space and Time with a Pre-existing closely allied Species"; and I showed how many peculiarities in the affinities, the succession, and the distribution of the forms of life, were explained by this hypothesis, and that no important facts contradicted it.

Even then, however, I had no conception of *how* or *why* each new form had come into existence with all its beautiful adaptations to its special mode of life; and though the subject was continually being pondered over, no light came to me till three years later (February, 1858), under somewhat peculiar circumstances. I was then living at Ternate in the Moluccas, and was suffering from a rather severe attack of intermittent fever, which prostrated me for several hours every day during the cold and succeeding hot fits. During one of these fits, while again considering the problem of the origin of species, something led me to think of Malthus' Essay on Population (which I had read about ten years before), and the "positive checks"—war, disease, famine, accidents, etc.—which he adduced as keeping all savage populations nearly stationary. It then occurred to me that these checks must also act upon animals, and keep down their numbers; and as they increase so much faster than man does, while their numbers are always very nearly or quite stationary, it was clear that these checks in their case must be far more powerful, since a number equal to the whole increase must be cut off by them every year. While vaguely thinking how this would affect any species, there suddenly flashed upon me the idea of *the survival of the fittest*—that the individuals removed by these checks must be, on the whole, *inferior* to those that survived. Then, considering the *variations* continually occurring in every fresh generation of animals or plants, and the changes of climate, of food, of enemies always in progress, the whole method of specific modification became clear to me, and in the two hours of my fit I had thought out the main points of the theory. That same evening I sketched out the draft of a paper; in the two succeeding evenings I wrote it out, and sent it by the next post to Mr. Darwin.[1] I fully expected it would be as new to him as it was to myself, because he had informed me by letter that he was

[1] These two papers are reprinted in my *Natural Selection and Tropical Nature.*

engaged on a work intended to show in what way species and varieties differ from each other, adding, "my work will not fix or settle anything." I was therefore surprised to find that he had really arrived at the very same theory as mine long before (in 1844), had worked it out in considerable detail, and had shown the MSS. to Sir Charles Lyell and Sir Joseph Hooker; and on their recommendation my paper and sufficient extracts from his MSS. work were read at a meeting of the Linnean Society in July of the same year, when the theory of Natural Selection or survival of the fittest, was first made known to the world. But it received little attention till Darwin's great and epoch-making book appeared at the end of the following year.

We may best attain to some estimate of the greatness and completeness of Darwin's work, by considering the vast change in educated public opinion which it rapidly and permanently effected. What that opinion was before it appeared is shown by the fact that neither Lamarck, nor Herbert Spencer, nor the author of the *Vestiges*, had been able to make any impression upon it. The very idea of progressive development of species from other species, was held to be a "heresy" by such great and liberal-minded men as Sir John Herschel and Sir Charles Lyell; the latter writer declaring, in the earlier editions of his great work, that the facts of geology were "fatal to the theory of progressive development." The whole literary and scientific worlds were violently opposed to all such theories, and altogether disbelieved in the possibility of establishing them. It had been so long the custom to treat species as special creations, and the mode of their creation as "the mystery of mysteries," that it had come to be considered not only presumptuous, but almost impious, for any individual to profess to have lifted the veil from what was held to be the greatest and most mysterious of Nature's secrets.

But what is the state of educated literary and scientific opinion at the present day? Evolution is now universally accepted as a demonstrated principle, and not one single writer of the slightest eminence, that I am aware of, declares his disbelief in it. This is, of course, partly due to the colossal work of Herbert Spencer; but for one reader of his works there are probably ten of Darwin's, and the establishment of the theory of the *Origin of Species by Means of Natural Selection* is wholly Darwin's work. That book, together with those which succeeded it, has so firmly established the doctrine of progressive development of species by the ordinary processes of multiplication and variation, that there is now, I believe, scarcely a single living naturalist who doubts it. What was a "great heresy" to Sir John Herschel in 1845, and "the mystery of mysteries" down to the date of Darwin's book, is now the common knowledge of every clever schoolboy, and of every one who reads even the newspapers. The only thing discussed now is, not the fact of evolution,—that is admitted—but merely whether or no the causes alleged by Darwin are themselves sufficient to explain evolution of species, or require to be supplemented by other causes, known or unknown. Probably so complete a change of educated opinion on a question of such vast difficulty and complexity, was never before effected in so short a time. It not only places the name of Darwin on a level with that of Newton, but his work will always be considered as one of the greatest, if not the very greatest, of the scientific achievements of the Nineteenth Century, rich as that century has been in great discoveries in every department of physical science.

## CHAPTER XIV

### POPULAR DISCOVERIES IN PHYSIOLOGY

Recluse, th' interior sap and vapour dwells,
In nice transparence of minutest cells..

—*H. Brooke.*

But a heavenly sleep
That did suddenly steep
In balm my bosom's pain.

*Shelley.*

THE science of Physiology, which investigates the complex phenomena of the motions, sensations, growth and development of organisms, is almost wholly the product of the present century; but with the exception of a few fundamental conceptions, it has been an almost continuous growth by small increments, and offers few salient points of popular interest, or which can be made intelligible to the general reader.

The first of the great fundamental conceptions referred to, is the cell-theory, which was definitely established for plants in 1838, and immediately afterwards for animal-structures. The theory is, that all the parts and tissues of plants and animals are built up of cells, modified in form and function in an infinite variety of ways, but to be traced in the early stages of growth, alike of bone and muscle, nerve and bloodvessel, skin and hair, root, wood, and flower. And, further, that all organisms originate in simple cells, which are almost identical in form and structure, and which thus constitute the fundamental unit of all living things.

The second great generalisation is what has been termed the recapitulation theory of development. Every animal or plant begins its existence as a cell, which develops by a process of repeated fission and growth into the perfect form. But if we trace the different types backward, we find that we come to a stage when the embryos of all the members of an order, such as the various species of Ruminants are undistinguishable; earlier still all the members of a class, such as the Mammalia, are equally alike, so that the embryos of a sheep and a tiger would be almost identical; earlier still all vertebrates, a lizard, a bird, and a monkey, are equally undistinguishable. Thus in its progress from the cell to the perfect form every animal recapitulates, as it were, the lower forms upon its line of descent, thus affording one of the strongest indirect proofs of the theory of evolution. The earliest definite result of cell-division is to form what is termed the "gastrula," which is a sack with a narrow mouth, formed of two layers of cells. All the higher animals without exception, from mollusc to man, go through this "gastrula" stage, which again indicates that all are descended from a common ancestral form of this general type.

One other physiological discovery is worth noting here, both on account of its remarkable nature and because it leads to some important conclusions in relation to the zymotic diseases. Quite recently it has been proved that the white corpuscles of the blood, whose function was previously unknown, are really independent living organisms. They are produced in large numbers by the spleen, an organ which has long been a puzzle to physiologists, but whose function and importance to the organism seem to be now made clear. They are much smaller and less numerous than the red blood-globules; they move about quite independently; and they behave in a manner which shows that they are closely allied to, if not identical with, the amœbæ found abundantly in stagnant water, and which form such interesting microscopic objects. These minute animal organisms, which inhabit not only our blood-vessels but all the tissues of the body, have an important function to perform on which our very lives depend. This function is, to devour and destroy the bacteria or germs of disease which may gain an entrance to our blood or tissues, and which, when their increase is unchecked, produce various disorders and even death. Under the higher powers of the microscope the leucocytes, as they are termed, can be observed continually moving about, and on coming in contact with any of these bacteria or their germs, or other hurtful substances, they send out pseudopodia from their protoplasm which envelopes the germ and soon causes it to disappear; but they also appear sometimes to produce a secretion which is injurious to the bacteria, and so destroys them, and these may perhaps be distinct organisms.

It seems probable, and, in fact, almost certain, that so long as we live in tolerably healthy conditions, these leucocytes (or phagocytes as they are sometimes called from their function of devouring injurious germs) are able to deal with all disease-germs which can gain access to our system; but, when we live in impure air, or drink impure water, or feed upon unwholesome food, our system becomes enfeebled, and our guardian leucocytes are unable to destroy the disease-germs that gain access to our organism; they then increase rapidly, and are in many cases able to destroy us.

We learn from this marvellous discovery, that, so long as we live simply and naturally, and obey the well-known laws of sanitation, so as to secure a healthy condition of the body, the more dreaded zymotic diseases will be powerless against us. But if we neglect these laws of health, or allow of conditions which compel large bodies of our fellow-men to neglect them, these disease-germs will be present in such quantities in the air and the water around us, that even those who personally live comparatively wholesome lives will not always escape them.

We learn, too, another lesson from this latest discovery of the secrets of the living universe. Just as we saw how, physically, dust was so important that not only much of the beauty of nature but the very habitability of our globe depended upon it, so we now find that the most minute and most abundant of all organisms, are those on which both our means of life and our preservation from death are dependent. For these minute bacteria of various kinds are present everywhere—in the air, in the water, in the soil under our feet. Their function appears to be to break up by putrefactive processes all dead organised matter, and thus prepare it for being again assimilated by plants, so as to form food for animals and for man; and it seems probable that they prepare the soil itself for plant-growth by absorbing and fixing the nitrogen of the atmosphere. They are, in fact, omnipresent, and under normal conditions they are wholly beneficial. It is we ourselves who, by our crowded cities, our polluted streams, and our unnatural and unwholesome lives, enable them to exert their disease-creating powers.

A brief notice must also be given of two discoveries in practical physiology, which have perhaps done more to benefit mankind that those great mechanical inventions and philosophical theories which receive more general admiration. These are, the use of anæsthetics in surgical operations, and the antiseptic treatment of wounds.

Anæsthetics were first used in dentistry in 1846, the agent being ether; while chloroform, for more severe surgical operations, was adopted in 1848; and though their primary effect is only to abolish pain, they get rid of so much nervous irritation as greatly to aid in the subsequent recovery. The use of anæsthetics thus renders it possible for many operations to be safely performed which, without it, would endanger life by mere shock to the system; while to the operating surgeon it gives confidence, and enables him to work more deliberately and carefully from the knowledge that the longer time occupied will not increase the suffering of the patient or render his recovery less probable. Nitrous-oxide gas is now chiefly used in dentistry or very short operations, sulphuric ether for those of moderate length, while chloroform is usually employed in all the more severe cases, since the patient can by its use be kept in a state of insensibility for an hour or even longer. There is, however, some danger in its use to persons with weak heart or of great nervous sensibility, and the patient in such cases may die from the effects of the anæsthetic alone.[1]

[1] The Hyderabad Chloroform Commission, which in 1889 thoroughly investigated the causes of death under chloroform, has proved that *all such deaths are preventible*, if a different

Even more important was the introduction of the antiseptic treatment in 1865, which, by preventing the suppuration of incised or wounded surfaces, has reduced the death-rate for serious amputations from forty-five per cent. to twelve per cent., and has besides rendered possible numbers of operations which would have been certainly fatal under the old system. I remember my astonishment when, soon after the introduction of the practice, I was told by an eminent physiologist of the new method of performing operations, in which the freshly cut surfaces could be left exposed to the air without dressings of any kind, and would soon heal. The antiseptic treatment was the logical outcome of the proof, that suppuration of wounds and all processes of fermentation and putrefaction, were not due to normal changes either in living or dead tissues, but were produced by the growth and the rapid multiplication of minute organisms, especially of those low fungoid groups termed Bacteria. If, therefore, we can adopt measures to keep away or destroy these organisms and their germs, or in any way prevent their increase, injured

mode of administration is adopted. And its conclusions have been confirmed by the independent researches of four medical men, —two English and two American physicians. Yet the old method of adminstration is still common in this country, no less than 75 deaths having occurred from this cause in 1896, while the Registrar General records 78 deaths from Anæsthetics (almost all from chloroform) in 1895. There is thus a terrible amount of mortality due, apparently, to the ignorance of medical men on a subject as to which they are supposed to have exclusive knowledge. An excellent account of the work of the above-named Commission is given in the "Nineteenth Century" of March 1898 by a lady who has had to take chloroform more than once, by both methods; and can therefore judge of their comparative effects by the best of tests—personal experience.

living tissues will rapidly heal, while dead animal matter can be preserved unchanged almost indefinitely. In the case of wounds and surgical operations this is effected by means of a weak solution of corrosive-sublimate, in which all instruments and everything that comes in contact with the wound are washed, and by filling the air around the part operated on with a copious spray of carbolic acid. Cold has a similar effect in preserving meat; while the process of tinning various kinds of food depends for its success on the same principle, of first killing all bacteria or other germs by heating the filled tins above the boiling point, and then keeping out fresh germs by air-tight fastening.

The combined use of anæsthetics and antiseptics has almost robbed the surgeon's knife of its terrors, and has enabled the most deeply-seated organs to be laid open and operated upon with success. As a result, more lives are probably now saved by surgery than by any other branch of medicine, since in the treatment of disease there has been comparatively small progress except by trusting more to the healing powers of nature, aided by rest, warmth, pure air, wholesome food, and as few drugs as possible.

## CHAPTER XV

### ESTIMATE OF ACHIEVEMENTS:—THE NINETEENTH AS COMPARED WITH EARLIER CENTURIES

The long crude efforts of society
  In feeble light by feeble reason led,—
But gleaning, gathering still, effect of cause,
  Cause of effect, in ceaseless sequence fed;
Till, slow developing the eons through,
  The gibbering savage to a Darwin grew—
This hath Time witnessed! Shall his records now,
The goal attain'd—the end achieved, avow?

—*J. H. Dell.*

HAVING now completed our sketch of those practical discoveries and striking generalizations of science, which have in so many respects changed the outward forms of our civilization, and will ever render memorable the century now so near its close, we are in a position to sum up its achievements, and compare them with what has gone before.

Taking first those inventions and practical applications of science which are perfectly new departures, and which have also so rapidly developed as to have profoundly affected many of our habits, and even our thoughts and our language, we find them to be thirteen in number.

1. Railways, which have revolutionized land-travel and the distribution of commodities.

2. Steam-navigation, which has done the same thing

for ocean travel, and has besides led to the entire reconstruction of the navies of the world.

3. Electric Telegraphs, which have produced an even greater revolution in the communication of thought.

4. The Telephone, which transmits, or rather reproduces, the voice of the speaker at a distance.

5. Friction Matches, which have revolutionized the modes of obtaining fire.

6. Gas-lighting, which enormously improved outdoor and other illumination.

7. Electric-lighting, another advance, now threatening to supersede gas.

8. Photography, an art which is to the external forms of nature what printing is to thought.

9. The Phonograph, which preserves and reproduces sounds as photography preserves and reproduces forms.

10. The Röntgen Rays, which render many opaque objects transparent, and open up a new world to photography.

11. Spectrum-analysis, which so greatly extends our knowledge of the universe, that by its assistance we are able to ascertain the relative heat and chemical constitution of the stars, and ascertain the existence, and measure the rate of motion, of stellar bodies which are entirely invisible.

12. The use of Anæsthetics, rendering the most severe surgical operations painless.

13. The use of Antiseptics in surgical operations, which has still further extended the means of saving life.

Now, if we ask what inventions comparable with these were made during the previous (eighteenth) century, it seems at first doubtful whether there were any. But we may perhaps admit the development of the steam-engine from the rude but still useful machine of Newcomen, to the powerful and economical engines of Boulton and Watt. The principle, however, was known long before, and had been practically applied in the previous century by the Marquis of Worcester and by Savery; and the improvements made by Watt, though very important, had a very limited result. The engines made were almost wholly used in pumping the water out of deep mines, and the bulk of the population knew no more of them, nor derived any more direct benefit from them, than if they had not existed.

In the seventeenth century, the one great and far-reaching invention was that of the Telescope, which, in its immediate results of extending our knowledge of the universe and giving possibilities of future knowledge not yet exhausted, may rank with spectrum-analysis in our own era. The Barometer and Thermometer are minor discoveries.

In the sixteenth century we have no invention of the first rank, but in the fifteenth we have printing.

The Mariner's Compass was invented early in the fourteenth century, and was of great importance in rendering ocean navigation possible and thus facilitating the discovery of America.

Then, backward to the dawn of history, or rather to prehistoric times, we have the two great engines of knowledge and discovery—the Indian or Arabic numerals, leading to arithmetic and algebra, and, more remote still, the invention of alphabetical writing.

Summing these up, we find only five inventions of

the first rank in all preceding time—the telescope, the printing-press, the mariner's compass, Arabic numerals, and alphabetical writing, to which we may add the steam-engine and the barometer, making seven in all, as against thirteen in our single century.

Coming now to the theoretical discoveries of our time, which have extended our knowledge or widened our conceptions of the universe, we find them to be about equal in number, as follows:—

1. The determination of the mechanical equivalent of heat, leading to the great principle of the Conservation of Energy.
2. The Molecular theory of gases.
3. The mode of direct measurement of the Velocity of Light, and the experimental proof of the Earth's Rotation. These are put together, because hardly sufficient alone.
4. The discovery of the function of Dust in nature.
5. The theory of definite and multiple proportions in Chemistry.
6. The nature of Meteors and Comets, leading to the Meteoritic theory of the Universe.
7. The proof of the Glacial Epoch, its vast extent, and its effects upon the earth's surface.
8. The proof of the great Antiquity of Man.
9. The establishment of the theory of Organic Evolution.
10. The Cell theory and the Recapitulation theory in Embryology.
11. The Germ theory of the Zymotic diseases.
12. The discovery of the nature and function of the White Blood-corpuscles.

Turning to the past, in the eighteenth century we may perhaps claim two groups of discoveries:—

1. The foundation of modern Chemistry by Black, Cavendish, Priestley, and Lavoisier; and

2. The foundation of Electrical science by Franklin, Galvani, and Volta.

The seventeenth century is richer in epoch-making discoveries, since we have:—

3. The theory of Gravitation established.

4. The discovery of Kepler's Laws.

5. The invention of Fluxions and the Differential Calculus.

6. Harvey's proof of the circulation of the Blood.

7. Roemer's proof of finite velocity of Light by Jupiter's satellites.

Then, going backward, we can find nothing of the first rank except Euclid's wonderful system of Geometry, derived from earlier Greek and Egyptian sources, and perhaps the most remarkable mental product of the earliest civilizations; to which we may add the introduction of Arabic numerals, and the use of the Alphabet. Thus in all past history we find only eight theories or principles antecedent to the nineteenth century as compared with twelve during that century. It will be well now to give comparative lists of the great inventions and discoveries of the two eras, adding a few others to those above enumerated.

| Of the Nineteenth Century. | Of All Preceding Ages. |
|---|---|
| 1. Railways. | 1. The Mariner's Compass. |
| 2. Steam-ships. | 2. The Steam Engine. |
| 3. Electric Telegraphs. | 3. The Telescope. |
| 4. The Telephone. | 4. The Barometer and Thermometer. |
| 5. Lucifer Matches. | 5. Printing. |
| 6. Gas illumination. | 6. Arabic numerals. |
| 7. Electric lighting. | 7. Alphabetical writing. |
| 8. Photography. | |

9. The Phonograph.
10. Röntgen Rays.
11. Spectrum-analysis.
12. Anæsthetics.
13. Antiseptic Surgery.
14. Conservation of Energy.
15. Molecular theory of Gases.
16. Velocity of Light directly measured, and Earth's Rotation experimentally shown.
17. The uses of Dust.
18. Chemistry, definite proportions.
19. Meteors and the Meteoritic Theory.
20. The Glacial Epoch.
21. The Antiquity of Man.
22. Organic Evolution established.
23. Cell theory and Embryology.
24. Germ theory of disease, and the function of the Leucocytes.

8. Modern Chemistry founded.
9. Electric science founded.
10. Gravitation established.
11. Kepler's Laws.
12. The Differential Calculus.
13. The circulation of the blood.
14. Light proved to have finite velocity.
15. The development of Geometry.

Of course these numbers are not absolute. Either series may be increased or diminished by taking account of other discoveries as of equal importance, or by striking out some which may be considered as below the grade of an important or epoch-making step in science or civilization. But the difference between the two lists is so large, that probably no competent judge would bring them to an equality. Again, it is noteworthy that nothing like a regular gradation is perceptible during the last three or four centuries. The eighteenth century, instead of showing some approximation to the wealth of discovery in our own age, is less remarkable than the seventeenth, having only about half the number of really great advances.

It appears then, that the statement in my first chapter, that to get any adequate comparison with the Nineteenth Century we must take, not any preceding century or group of centuries, but rather the whole preceding epoch of human history, is justified, and more than justified, by the comparative lists now given. And if we take into consideration the change effected in science, in the arts, in all the possibilities of human intercourse, and in the extension of our knowledge, both of our earth and of the whole visible universe, the difference shown by the mere numbers of these advances will have to be considerably increased on account of the marvellous character and vast possibilities of further development of many of our recent discoveries. Both as regards the number and the quality of its onward advances, the age in which we live fully merits the title I have ventured to give it of—THE WONDERFUL CENTURY.

PROGRESS!

Not empanoplied as Pallas, with her spear and Gorgon
shield,
But with fair Athene's olive, peaceful Progress takes the
field;
Yet that shield is ever ready, and that spear is hers at
need,
To protect the field she cultures, and defend the garnered
seed;
And the meanest in her legions, marching with a level
breast
In unbroken line of duty with her bravest and her best,
Answering only to her watchwords, walking only by her
light,
Mustering to her only banner—to the gonfalon of Right;
For that flag's unstainèd honour—in that flag's unswerving
cause—
Knows no other teacher's credo—owns no other leader's
laws;
Treads that only Temple's pavement by the feet of Reason
trod,
That hath Truth alone for Priestess—Equity alone for
God.

—*J. H. Dell.*

## RIGHT!

Thy cause is Right, gird then thy loins her mandate to fulfil;
Unswerving purpose, fix'd resolve, indomitable will—
Save these ask no auxiliar arm, but cause-reliant stand,
Although the foe be myriads strong, and thine a single brand;
Have thou but faith, firm faith alone, and, though the world assail,
The will-drawn sword that Justice girds shall 'gainst all odds prevail.
That charmèd sword, nor foe can wrench, nor enemy can wield,
It may not fall to adverse hand, the spoil of adverse field,
With feint, nor guile, smirch thou its blade, but forward boldly dare,
And bear thou thence Right's victory, or leave her champion there.

*—J. H. Dell.*

# Part II—Failures

## CHAPTER XVI

### THE NEGLECT OF PHRENOLOGY

All be turned to barnacles, or to apes
With foreheads villainous low.

—*Shakespeare.*

His searching wisdom taught
How the high dome of thought
Pictured the mind;
On that fair chart confest,
Traced he each reckless guest
Which in the human breast
Lies deep enshrined.

—*Eulogy of Dr. Gall.*

In the preceding chapters I have, to the best of my ability, given a short, but I trust accurate, sketch of the most prominent examples of material and intellectual progress during the Nineteenth Century. In doing this I have fully recognized the marvellous character of many of these discoveries, as well as the great amount, and striking novelty, of the material advances to which they have given rise. But, along with this continuous progress in science, in the arts, and in wealth-production—which has dazzled our imaginations to such an extent that we can hardly admit the possibility of any serious evils having accompanied or been caused by it—there have been many serious failures, intellectual, social, and moral. Some of our great thinkers have been so impressed by the terrible nature of these failures, that they have doubted whether the final result of the work of the century has any balance of good over evil—of happiness over misery, for mankind at large. But although this may be an exaggerated and pessimistic view, there can be no doubt of the magnitude of the evils that have grown up or persisted, in the midst of all our triumphs over natural forces, and our unprecedented growth in wealth and luxury.

We have also neglected or rejected some important lines of investigation affecting our own intellectual and spiritual nature; and have in consequence made serious mistakes in our modes of education, in our treatment of mental and physical disease, and in our dealings with criminals. A sketch of these various failures will now be given, and will, I believe, constitute not the least important portion of my work. I begin with the subject of Phrenology, a science of whose substantial truth and vast importance I have no more doubt, than I have of the value and importance of any of the great intellectual advances already recorded.

In the last years of the eighteenth century Dr. François Joseph Gall, a German physician, discovered (or rediscovered) the facts, now universally admitted, that the brain is the organ of the mind, that different parts of the brain are connected with different mental and physical manifestations, and that, other things being equal, size of the brain and of its various parts is an indication of mental power. He began his observations on this subject when a boy, by noticing the different characters and talents of his schoolfellows—some were peaceable, some quarrelsome; some were expert in penmanship, others in arithmetic; some could learn by rote even without comprehension, while others, although more intelligent, could not do so. He himself was one of the latter group; and this led him to notice that those who surpassed him most in this power of verbal memory, however different they might be in other respects, had all prominent eyes. The meaning of this peculiarity he did not at the time perceive, but he continued his observations at college and in the hospitals, and very gradually acquired the certainty that strongly marked peculiarities of character or talent were associated with constant peculiarities in the form of the head. This led him to pay special attention to the anatomy of the brain and its bony covering; he made collections of skulls and casts of skulls of persons having special mental characteristics; he collected also the skulls of various animals, and compared their brains with those of man; he visited prisons, schools, and colleges, everywhere making observations and comparisons of form and size with mental faculties; and later on, when he became physician to a lunatic asylum in Vienna, he had vast opportunities for studying the diseased brain, and for observing the correspondence between the form of the head and the special delusions of each patient.

It was after more than twenty years of continuous observation and study, under exceptionally favourable conditions, that he became convinced that he had discovered a real connection between the mental faculties and the form and size of the various parts of the brain; and in the year 1796 he began lecturing on the subject. His lectures were continued for five years, and were attended by numerous physicians and medical students, as well as by men of culture of all ranks, many converts being made. The lectures were then forbidden by the authorities, on the ground that he had not had permission to deliver them. He declined to ask for permission, and soon afterwards left Vienna, and with his most distinguished pupil, Dr. Spurzheim, travelled through a large part of Northern Europe, lecturing in the chief cities, and finally settled in Paris in 1807. In 1813 Spurzheim visited Great Britain, where he lectured for four years; and it was during this period that George Combe made his acquaintance in Edinburgh, and thenceforth began that long course of personal observation and study which rendered him the best English exponent of the science, and probably one of the best practical phrenologists of any country.

Combe was a man of great mental power, extremely logical, ardent in the pursuit of truth, but also extremely cautious in ascertaining what was and what was not true. A clever writer in the *Edinburgh Review*—Dr. John Gordon—had just condemned and ridiculed the doctrines of Gall and Spurzheim as being full of absurdities and misstatements, and "a piece of thorough quackery from beginning to end." It was a clever and vigorous critique, apparently founded on knowledge; and Combe read it with so much enjoyment and conviction, that when shortly afterwards Spurzheim came to Edinburgh and gave a course of lectures, he refused to go and hear him. When the lectures were over, however, a friend asked Combe if he would like to come to his house and see Dr. Spurzheim dissect a brain; and as he was always eager for knowledge, and had already studied anatomy, he went. Combe had been a physiological student under Dr. Barclay, and had often seen him dissect the brain, but was taught nothing of its functions, of which the lecturer had declared that nothing was known. But when Dr. Spurzheim dissected, Combe tells us that he at once saw how "inexpressibly superior" was his method, in showing its detailed structure; while he saw at the same time that the reviewer had displayed profound ignorance, and had been guilty of gross misrepresentation. He therefore attended Spurzheim's second course of lectures, and was so impressed that he determined to observe and study for himself. He at once ordered from London a collection of casts of the skulls of men of known mental peculiarities—artists, writers, workers, criminals, etc.; but when they arrived, the differences looked so slight that he thought he should never be able to determine the peculiarities which, on Dr. Spurzheim's theory, were so important, and therefore determined to put them aside and trouble no more about them. But their arrival was known to some of his friends, and numbers of persons called asking to see them, and begging him to explain their phrenological peculiarities. He was thus *forced* to observe them more carefully; and as he showed them to each fresh visitor he began to see that there were large differences between them, and that these differences corresponded to the differences of their known characters according to the position of the organs as determined by Gall and Spurzheim. He thus obtained confidence in his powers of observation, and therefore determined to go on with the study. He began to observe the heads of all his friends and clients, and found that these usually confirmed the experience already gained. This gave him confidence; and for three years he went on studying both the heads of living persons and actual crania, the latter more especially, in order to learn the exact amount of correspondence or difference between the outer and inner surfaces of the skull. His visitors increased as his knowledge rendered his explanations more interesting, and thus, he tells us, he became a phrenologist and a lecturer on phrenology by a concatenation of circumstances which were not foreseen, and the ultimate consequences of which he had never contemplated.

Before proceeding further with a sketch of the evidences for phrenology, it is well to consider briefly what sort of man Combe was. At the period just referred to he was twenty-seven years old, and in good practice in Edinburgh as a lawyer. He carried on his profession for twenty years longer, his practice continually increasing, notwithstanding his various other occupations and the unpopularity of many of his writings. During this time he had written and published several works—some very extensive—on *Phrenology; The Constitution of Man*—a work which in Scotland caused him to be considered an infidel, but which in England had a circulation of a hundred thousand; *Lectures on Popular Education*; *Lectures on Moral Philosophy*, afterwards enlarged into a work which went through several editions, besides numerous articles in periodicals and newspapers on a variety of subjects. Though brought up in a religious Scotch family, and of a highly reverential nature, he entirely emancipated himself from religious dogmas, and became the best exponent of a well-reasoned system of natural religion. He was one of the earliest educational reformers, and may almost be considered as the founder of rational systems of education in this country. Wherever he went—and he visited repeatedly many European countries as well as the United States—his great reputation as a religious, social, and educational reformer and philosophical thinker, led to his being welcomed in the best social, scientific, and political circles. At home he was consulted by many persons of eminence, including the Prince Consort, on the best system of education for their children. Sir James Clark, Richard Cobden, Robert Chambers, and Charles Mackay the poet, were among his intimate friends; while Lord John Russell and other influential politicians, were glad to receive information from him on all subjects connected with improved systems of education.

It may be truly said that on every subject on which he wrote—the constitution of man, natural religion, education, criminal legislation, the lunacy laws, the currency question, moral philosophy—he was far in advance of his age; and almost all his principles and his proposals on these subjects, though considered heretical or impracticable by most of his contemporaries, are now either actually adopted or admitted to be correct both in philosophy and in practice. But the one subject to which he gave more careful study than to any other—phrenology—which was indeed the very foundation on which his philosophy and his educational theories were built, was contemptuously rejected by the great bulk of the scientific and literary men of his time, without adequate examination, without any reasonable study of so complex and important a subject, but almost entirely on false assumptions, gross misrepresentations, or *a priori* reasoning. All who have given any careful consideration to the writings of Dr. Gall and George Combe, admit, that both were men of exceptional mental power, careful observers, close reasoners, cautious in arriving at conclusions on anything less than overwhelming evidence. The first gave all his energies during a long life to the establishment, on a firm basis of observation and experiment, of the new science of Phrenology which he had founded; the second, coming to the subject with prepossessions against it, took nothing for granted, observed every alleged fact for himself, criticised, modified and extended the work of his teachers, and taught it by lectures and books in a manner at once popular and scientifically exact. And the life-work of two such men was disposed of, not by pointing out important errors of observation or of reasoning, but largely by abuse, or by means of trivial objections which the most rudimentary knowledge shows to be unfounded.

Let us now consider, briefly, what phrenology is, what is the evidence on which it is founded, and what are its practical results. In the first place it is a purely inductive science, founded step by step on the observation and comparison of facts, confirmed and checked in every conceivable way, and subjected to the most rigid tests. By means of large collections of skulls, and casts of the heads of men and women remarkable for any mental faculty or propensity, and by observations and measurements of thousands of living persons, the correspondence of form with function was first suspected, then confirmed, and finally demonstrated by the comparison of the heads of individuals of every age, both in health and disease, and under the most varied conditions of education and environment. Three men of exceptional talents and acuteness of observation devoted their lives to the collection of these facts. They studied also the brain itself, and discovered many details of its structure before unknown. They studied the skull, its varying thicknesses in different parts and at different ages, as well as under the influence of disease. And it was only after making allowance for every source of uncertainty or error that they announced the possibility of determining character with a considerable amount of certainty, and often with marvellous exactness. Surely this was a scientific mode of procedure, and the only sound method of ascertaining the relations that exist between the development of the brain and the mental faculties and powers. A few examples showing how far this was actually done will now be given.

In October, 1835, Combe visited the Newcastle Lunatic Asylum and examined the heads of several of the patients. These were selected by the Surgeon-Superintendent, Mr. Mackintosh, and their mental peculiarities had been noted down by him beforehand. For convenience of comparison, Combe's notes and those of Mr. Mackintosh are put in parallel columns.[1]

| *Combe's Phrenological Notes.* | *Superintendent's Notes.* |
|---|---|
| PATIENT J. N. | |
| ANIMAL organs large. | A bad character. |
| CAUTIOUSNESS and DESTRUCTIVENESS predominant. | Hypochondriacal. |
| HOPE small. MORAL FACULTIES deficient. | Suicidal. |

[1] These tests at Newcastle are fully reported in the *Phrenological Journal*, vol. ix. pp. 519–526.

PATIENT L. J.

| | |
|---|---|
| ACQUISITIVENESS enormously large. | Monomania, wealth. |

PATIENT J. M.

| | |
|---|---|
| Intellectual organs well developed. | Generally sane and tractable. |
| VENERATION, CONCENTRATIVENESS *very* large. | Monomania, the Messiah. |
| FIRMNESS, SELF-ESTEEM large. | A proselyte Jew; will lead the Jews to the conquest of England. |

PATIENT C. S.

| | |
|---|---|
| Intellectual organs large. | |
| Organ of NUMBER *exceedingly* large. | Dementia — perpetually employed with figures and arithmetic. |

FEMALE PATIENT M. D.

| | |
|---|---|
| Moral faculties deficient. | |
| HOPE extremely small. | Great misery. |
| DESTRUCTIVENESS and CAUTIOUSNESS excessively large. | Suicidal monomania. |

At the Dunstane Lodge Asylum, near Newcastle, Mr. Combe, attended by two surgeons, the editor of the *Tyne Mercury*, and a few other gentlemen, examined the heads of a few patients submitted to him by the proprietor, Mr. Wilkinson, who appended his own remarks on the nature of their insanity.

| *Mr. Combe's Delineation.* | *Mr. Wilkinson's Remarks.* |
|---|---|
| PATIENT J. F. | |
| SELF-ESTEEM and FIRMNESS very large. WONDER, SECRETIVENESS and ACQUISITIVENESS also large. The character of the insanity will be self-esteem, and probably cunning and theft. | He proclaims himself to be the Great God, and entertains a high esteem of his person and strength. He pilfers and picks up little articles whenever he can lay his hands on them. |

PATIENT R. M.

| | |
|---|---|
| Intellectual organs large.<br>IMITATION very large.<br>COMBATIVENESS and DESTRUCTIVENESS very large.<br>HOPE and CONSCIENTIOUSNESS deficient.<br>Character very violent; probably attempted suicide; great power of expressing his feelings by his countenance and gestures. | He has a talent for all kinds of mechanical work. He is extremely violent, and has a great talent for imitation. His countenance is fearfully expressive when he is excited. |

PATIENT H. C.

| | |
|---|---|
| Large COMBATIVENESS, enormous SELF-ESTEEM.<br>FIRMNESS and PHILOPROGENITIVENESS large.<br>Intellect and IMITATION large.<br>He will manifest extreme conceit with great determination. He will have a great talent for imitation and strong powers of natural language. | This exactly describes the character. He believes himself to be a king; he is prone to imitate; he is opinionative, and fond of children. |

On October 28th in the same year Mr. Combe visited the Newcastle Gaol, accompanied by several medical gentlemen and others who had attended his lectures. Several of the criminals were examined by him, and while he was writing down their characteristics, Dr. George Fife, the assistant-surgeon to the gaol, who knew nothing of phrenology, wrote a brief account of their characters from his personal knowledge. The following are the three cases submitted:—

| *Mr. Combe.* | *Dr. Fife.* |
| --- | --- |
| P. S. (aged 20). | |
| My inference is that this boy is not accused of *violence*; he has a talent for deception and a desire for property not regulated by justice. It is most probable that he has *swindled*: he has the combination which contributes to the talent of an *actor*. | Twice convicted of theft. He has never shown *brutality*, but he has no sense of honesty. He has frequently attempted to *impose* on Dr. Fife. . . . He has a talent for *imitation*. |
| T. S. (aged 18). | |
| This boy is very different from the last. He has probably been committed for assault connected with women. He may have stolen, though I think this less probable. He has fair intellectual talents, and is *improvable*. | Crime, rape. . . . Mild disposition; has never shown actual vice. |
| J. W. (aged 73). | |
| Case for a lunatic asylum rather than a gaol. Moral organs *very* defective. Intellect moderate. Cautiousness very large. No control of the lower propensities. | A thief; obstinate, ungrateful; one of the most depraved characters. |

Another interesting test-case is the following. A surgeon at Chatham sent a skull to Dr. Elliotson, stating that he belonged to a literary society the members of which were much divided on the subject of phrenology, and it was suggested that the skull in question, being that of a person whose character and previous history was known to the members, should be sent to some eminent phrenologist with a request for a delineation of the character. Dr. Elliotson, to whom the person who sent the skull was quite unknown, gave him the following sketch of the character of the deceased person :—

"I should say that he was a man of strong passions, which overbalanced his intellect; that he was prone to *great violence*, but *by no means courageous*; that he was *extremely cautious and sly*; his sexual desires were strong, but *his love of offspring very remarkable*. I can discover no good quality about him except the love of his children, if he had any. The most striking *intellectual* quality in him, I should think, was his *wit*. He might also have been a good *mimic*."

The actual history and character of the man are given at length, but the following are the main points. He was of respectable parentage, but was sensual and vicious. He became a farmer in Cheshire, and took to *smuggling salt*, which was then contraband; but he always escaped detection, though long suspected. Later, he made use of his assistants in the smuggling business for the purpose of robbing the farmers around of corn, which, being a farmer, he was able to sell *without suspicion*. He was at length detected and condemned to death, a sentence which was commuted to transportation for life. He was, however, on account of his age, not sent abroad, but kept in the convict-hulks. After two years, being in very bad health, he was transferred to the hospital-ship, where he remained till his death. Here he was very reserved as to his own history, but, being treated with great kindness he made statements to the following effect:—(1) That though he had led a lawless life he had never committed murder. (2) That he had a wife and eight children, a natural son in Wales, and that he had several mistresses in different parts of the country up to the time of his apprehension.

In the hospital he exhibited a severe *sarcastic wit* at the expense of those around him. The manners and language of the clergyman at the hospital were the frequent subjects of *his mimicry*. He exhibited a strong *attachment to his children*, and frequently spoke of them in the most affectionate manner.

It will be observed that *all* the special features of the man's character, as given by Dr. Elliotson, were strictly correct, although the combination was an uncommon and remarkable one; and every unprejudiced person will agree with the following resolution, which was passed by the Society unanimously, and transmitted to Dr. Elliotson:—

"That the character given of L. by Dr. Elliotson, from the inspection of the skull, corresponds so exactly with his history, that it is impossible to consider the coincidence as the effect of chance, but that it is an instance which, if supported by many others, affords a strong foundation for the truth of phrenology."[1]

One other test of a remarkable character is given in the same volume as that containing the above (p. 467). In the spring of 1826 Dr. Thomson, a Navy surgeon, had charge of 148 convicts on the voyage to New South Wales. A friend of the doctor's induced him to allow a phrenologist, Mr. De Ville, to make an examination of the whole number, giving the surgeon a memorandum, which he might compare with the actual character of the men during the long voyage. This was done; and one man in particular was noted as being "very dangerous from his energy, ferocity, and talent for plots and profound dissimula-

[1] The above experiment with correspondence is given in full in the *Phrenological Journal*, vol. iv. p. 258.

tion." The voyage occupied four months, and the surgeon kept a careful official journal as regards the convicts, the main facts of which are summarized in the following letter to his friend, dated Sydney, October 9th, 1826:—

"I have to thank you for your introduction to De Ville and to phrenology, which I am now convinced has a foundation in truth, and beg you will be kind enough to call on Dr. Burnett, whom I have requested to show you my Journal, at the end of which is Mr. De Ville's report, and my report of conduct during the voyage. . . *De Ville is right in every case except one*—Thomas Jones; but this man can neither read nor write, and, being a sailor, he was induced to join the conspiracy to rise and seize the ship and carry her to South America, being informed by Hughes, the ringleader, that he would then get his liberty. Observe how De Ville has hit the real character of Hughes, and I will be grateful to De Ville all my life; for his report enabled me to shut up in close custody the malcontents and arrive here not a head *minus*, which, without the report, it is more than probable I should have been. All the authorities here have become phrenologists, and I cannot get my journals out of their offices until they have perused and reperused De Ville's report." [1]

One more case only can here be given. Combe reviewed a volume by Archbishop Whately, which led to some correspondence, and the Archbishop sent Combe a cast of his head, asking for his unbiassed opinion. The Archbishop was much struck by the character-sketch sent, but wishing for a more complete test, requested Combe to send the cast to some other phren-

[1] See *Phrenological Journal*, vol. iv. p. 467.

ologist, *without any indication of the person it represented*, and let him know the result. Combe did so, and the resulting report was shown to two of the Archbishop's most intimate friends, who expressed their wonder at the accuracy with which the character had been unfolded, declaring that, except in a few minor details, they could find nothing to correct. The same cast was then sent to a third phrenologist, and the Archbishop gave the following personal details in reference to the two last:—"What I was most struck with was, in the one, my difficulty of withstanding solicitations; in the other, my delight in an infant-school. The former, though well known to myself, was, I believe, never detected in my conduct."

I will now briefly state my own experiences of phrenological delineation, the accuracy of which confirmed me in the belief that the science is a true and important one, which I had already reached by a study of the works of George and Andrew Combe. When I was about three or four and twenty, living at Neath, Glamorganshire, I had my head examined by two phrenological lecturers who visited the town at different times. As the fee for a full delineation was rather high I only received a sketch, and many details were therefore omitted. But all that was stated was correct, and much of it remarkably so, as shown by the following extracts:—

1. "You will pay great attention to facts, but so soon as facts are presented you will begin to reason and theorize upon them. You will be constantly searching for causes." 2. "You will be a good calculator, will excel in mathematics, and will be very systematic in your arrangements." 3. "You

possess a good deal of firmness in what you conceive to be right, but you want self confidence."

These are the main points of the least full and least successful delineation, and the only error is, that my mathematics are strictly *limited*, as indicated in the better delineation from which I extract the following :—

4. This gentleman should learn easily and remember well, notwithstanding *verbal* memory is but moderate." 5. "He has some vanity but more ambition. He may occasionally exhibit a want of *self-confidence*; but general opinion ascribes to him too much. In this, opinion is wrong. *He* knows that he has not enough." 6. "If *Wit* were larger he would be a good mathematician, but, without it, I do not put his mathematical abilities as first-rate." 7. "He has some love for music from his *Ideality*, but I do not find a good ear or sufficient *Time*." 8. "He is fond of argument and not easily convinced."

Nos. 1 and 8 combined with large *Ideality* and *Wonder* (as indicated by both phrenologists) giving a strong love of the beauties and the mysteries of nature, furnish the explanation of my whole scientific work and writings.

Nos. 2 and 6 are exceedingly suggestive on account of their curiously precise estimate of faculty. At school I was good at arithmetic and elementary algebra, which always had a fascination for me; but as I left school when only fourteen I did not advance far. After I came of age, however, I was for two years English and Drawing Master in the Collegiate School at Leicester, the Head Master of which was a high Cambridge Wrangler; and he kindly offered to assist me in the higher mathematics. I worked through *Hind's Equations* and *Trigonometry* successfully, got on with the *Differential Calculus* with some difficulty, but broke down over the *Integral Calculus*, for want of that faculty of intuitively perceiving resemblances and incongruities, whether in ideas, words or symbols, somewhat awkwardly termed by phrenologists "Wit," but defined by some as the "organ of analogy." As a fact, I have no power to joke or make a pun, or see quickly all the possibilities of a position in chess, though no one more enjoys these diversions than myself. Most great mathematicians are either witty or poetical—Rankine, Clifford, De Morgan, Clerk-Maxwell, and Sylvester being well known examples; and that a phrenologist should detect my failure in the higher mathematics, and connect it with the deficiency of this organ, has always seemed to me very remarkable.

Nos. 3 and 5 both dwell on my want of self-confidence, and the second says that I am often thought to have too much. This is very true. In youth I was painfully shy, and was literally *afraid* of calling on people without an invitation. When I was in Para, in 1848, I was accused of being too proud to call on people, and suffered much in consequence; and throughout my whole life I have never been able to become intimate with any persons except those whose manners and dispositions were such as to make me at once feel sympathetic and at home with them. I have therefore made fewer intimate friends than most men, and all for want of a larger development of self-esteem.

No. 4 indicates my deficiency of verbal memory,

due to a small organ of Language. This makes the acquiring of foreign languages painful to me, and interferes with my success as a public speaker; since, though I know what ideas or arguments I wish to advance, I cannot at once find the right words by which to express them adequately, and in the effort to find the words the connection of ideas is liable to be lost.

Lastly, No. 7 states the exact nature of my mind in relation to music. Grand or pathetic music affects me strongly; but I should not detect considerable errors in the performance, my ear, as it is termed, being exceedingly deficient, while my perception of time is only a trifle better.

There are some other estimates as to my innermost nature which I know to be correct, but which are not suitable for exposition here; and these, combined with the more obvious characteristics above enumerated, produced a strong impression on my own mind as to the value of phrenology, which has remained unimpaired throughout my life.

The evidence of the value of phrenology in determining the hidden springs of character here given, might be increased ten or twenty-fold from the records of the early part of the century; and they produced an effect on the public mind which has not yet disappeared, since it is not an uncommon thing to meet with people who are quite unaware that the phrenology of their youth has been wholly rejected by the scientific world of to-day. Let us therefore now briefly consider how and why it was so rejected.

The first great objection was a religious one. The orthodox clergy both in Scotland and England held it to be contrary to Scripture and dangerous to morality. These objectors, of course, never made any pretence of studying the subject, or even of ascertaining what it really was. They decided at once that it was irreligious, and their flocks, for the most part, followed them.

The next body of opponents was that of the metaphysicians, headed by the great name of Sir William Hamilton. These philosophers, as they termed themselves, had from the earliest ages studied the mind by observations on their own consciousness, and on the mental operations of others so far as they could detect them. They recognised no connection between the mind and the organism; and as the phrenologists maintained that they had not only proved such a connection, but had also determined the particular parts of the brain which were the organs of the separate faculties—many of which the metaphysicians did not recognise at all—they of course declared the whole science to be erroneous, and its teachers to be little better than deluded fanatics. These objectors, also, never condescended to make any personal study of the science, and remained quite ignorant of its facts or of the mass of evidence which had been collected in support of it.

The third class of opponents consisted mainly of doctors and physiologists. At first, large numbers of these were converted by attending the lectures of Gall, Spurzheim, and Combe. In fact there is, so far as I can find, no record of any medical men or others who, having first attended a complete course of lectures, then proceeded to apply and test the information they had obtained with an earnest desire to ascertain the truth of the matter, who did not become confirmed phrenologists. Down to about the years 1840 or 1845 phrenology continued to progress, and there then seemed to be no reason why it should not take its place among the recognised sciences, since it was acknowledged by such men as Sir James Clarke, Physician to the Queen, Sir John Forbes, M.D., Dr. Elliotson, Dr. William Gregory, Dr. Engledue, Dr. Conolly, Physician of Hanwell Asylum, Dr. Abernethy, Professor of anatomy and surgery to the College of Surgeons, and many others. Soon after this period, however, it began to decline; and as the causes which led to this decline have, I believe, never been clearly pointed out, I will here state them as they seem to me to have acted.

The two main causes which discredited Phrenology appear to have been (1) the increase of itinerant lecturers, many of whom were uneducated, and some ignorant of the subject they professed to expound; and (2) its association with mesmerism or hypnotism, which at that time was still more virulently opposed.

1. Although phrenology, to be thoroughly understood and applied to the accurate delineation of character, requires a considerable amount of study and long practice, yet it appears, superficially, to be very easy; and it can actually be applied in cases of very marked character with fair success after a moderate amount of practice. Hence, although many of the public exponents of the science were very able men, there were others who adopted the business of lecturer and examiner of heads with imperfect knowledge. These, by their ignorance of the anatomy and physiology of the brain, their clumsiness in detecting the com-

parative size of the organs, and their inability to estimate the complicated results produced by the various combinations of the organs as influenced by temperament, education, and social position, were liable by their mistakes to bring great discredit on the subject, since the public, and especially those who opposed phrenology from any of the causes already stated, could not, or would not, distinguish between the student and the pretender, and loudly proclaimed that these failures demonstrated the fallacy of the whole science. Considering all these sources of opposition and disrepute, and the difficulty of moving the established sciences and professions, or the official world, to recognise any new thing, it is not to be wondered at, that when the enthusiasm of the early investigators and discoverers had passed away, no new students were found of sufficient independence, ability, and position to take their place.

2. Just about the time when Phrenology was gaining a wide acceptance, painless operations during the mesmeric trance were exciting the fiercest opposition of the medical profession; and Dr. John Elliotson, President of the Medical and Chirurgical Society, Lecturer at St. Thomas' Hospital, and a Professor at the University of London—an ardent phrenologist and founder of the Phrenological Association—was the chief defender of these painless operations, for supporting and practising which his professorship was taken from him. As regards this question of hypnotism, Dr. Elliotson is now known to have been right, and his opponents and traducers wholly wrong and grossly prejudiced, as will be shown in our next chapter; yet this prejudice undoubtedly reacted upon phrenology, and, together, with the theological and metaphysical

prejudice and that caused by imperfectly educated lecturers and professors, checked the official recognition it might otherwise have received, and rendered it impossible for students of medicine to become avowed phrenologists without injury to their professional prospects.

These combined influences led to its being treated as altogether a fallacy; and so complete became the ignorance of it among physiologists and medical men in the latter half of the century, that it was, and is, often spoken of as a purely fantastic scheme, the product of the *imaginations* of its founders, and entirely unsupported by observation and experiment. The complete ignorance of *how* phrenology was discovered by Gall, and of the enormous body of carefully observed facts and experiments it was founded upon, is well shown by the absurdly trivial nature of the objections made to it, even by men who might be supposed at least to have read some of the works of its founders before rejecting it. The most common and often-repeated objection is that of the frontal sinuses and the varying thickness of the skull in different parts and in different individuals, which are adduced as if they were known only to the objectors, and as if the eminent anatomist who devoted thirty years to the study of the brain and its bony covering had remained quite ignorant of them! If the objectors had read any work upon phrenology, they would have found that this was one of the very earliest of the small difficulties which the phrenologists recognised and overcame, and which every student learns how to allow for; while, if it were a much greater difficulty than it is, it could only affect the practical application of phrenology in certain cases and to a limited extent, without in any way disturbing its general principles or the vast body of facts on which it is founded. Even so eminent a physiologist and so careful a thinker as the late Professor Huxley, when I once asked him *why* he did not accept phrenology as a science, replied at once, "Because, owing to the varying thickness of the skull the form of the outside does not correspond to that of the brain itself, and therefore the comparative development of different parts of the brain cannot be determined by the form of the skull." To this I replied that the thickness of the skull varied at most by a few *tenths* of an inch, whereas the variations in the dimensions and the form of the head as measured in different diameters varied by whole *inches*, so that the size and proportions of the head as measured or estimated by phrenologists were very slightly affected by the different thicknesses of the skull, which, besides, had been carefully studied by phrenologists as dependent on temperament, age, etc., and could in many cases be estimated. He admitted the correctness of this statement, and had really no other objection to make, except by saying that he always understood it had been rejected after full examination (which it certainly had not been), and to ask, if it were true why was it not taught by any man of scientific reputation?

Almost the only other serious objection is to the detailed classification of the mental faculties, and to the names given to the several organs. But such objections exist even in the best established sciences, such as geology, where both classification and nomenclature are continually changing in the effort to approach nearer to the facts of nature. Phrenology is a science of observation as truly as is geology itself; it is a highly complex and difficult study, and it can hardly be supposed that the half-dozen eminent men who established it have exhausted its possibilities. The classification, or rather the enumeration of the mental faculties, whose function has been found to be dependent on certain brain-areas, is wholly founded on long-continued observation and comparison; and there is, of course, room for improvement, founded on further observations. But in this case, the objections of those who classify the mental faculties from their own consciousness are of no avail. Our consciousness does not reveal the brain-organs on which the faculties depend, and cannot therefore be used to criticise phrenology, which is the science of this dependence. And in like manner the older anatomists, who only *dissected* the brain, had no valid grounds of objection, since, as Combe always urged, "Dissection never reveals functions."

But while rejecting phrenology, neither anatomists, physiologists, nor anthropologists were able to give us any knowledge of the relations of mind and brain by other means. Enormous collections of skulls were formed; they were figured and accurately measured, were classified as brachycephalic or dolichocephalic, and in various other ways, but nothing came of it all, except a rough determination of the average size and typical form of skull of the different races of man with no attempt whatever to connect this typical form with the mental peculiarities of the several races. Never perhaps was so much laborious scientific work productive of so inadequate a result.

But about the year 1870 several continental physiologists, and in this country, Professor Ferrier, began to experiment on the brains of living animals, which were excited by weak galvanic currents applied to the exposed surface at different spots, and the resulting visible effects observed. In this way it was found that the excitement of certain limited areas caused the contraction of definite sets of muscles, leading to motion of the limbs, body, face, or head of the animal. This was termed the Localisation of Functions of the Brain, and was at once adduced by popular writers as giving the final death blow to phrenology, since it showed (as they ignorantly assumed) that portions of the brain which the phrenologists had alleged to be the organs of purely mental faculties were really only organs of muscular movements. Such writers entirely overlooked the very obvious considerations that the brain may be, in fact, must be, the centre for the production of movements as well as for initiating ideas; and that the rude method of exciting the living brain by galvanism, was not likely to develop the purely mental phenomena, which, indeed, in the animals experimented on, could only be exhibited *through* muscular movements. Again, it is quite possible, and even probable, that, while the cortex or grey matter on the surface of the brain is the seat of ideation, the more deeply seated matter may contain the centres for muscular and nervous action, and may be the part which is excited by the galvanic current. But this very fact of the connection of certain definite brain-areas with muscular motion is no new discovery, as modern writers seem to suppose, but was known to Dr. Gall himself, although he did not possess the modern appliances for the full experimental demonstration of it. In one of his first writings upon his discoveries—his letter to Baron de Retzer upon the Functions of the Brain in Man and Animals—he stated that there was a strange communication of the muscles with cerebral organs, adding—"when certain cerebral organs are put in action you are led, according to their seat, to take certain positions, as though you are drawn by a wire, so that one can discover the seat of the acting organs by the motions." This is the natural "expression of the emotions" which was so well studied by Darwin, but which Gall at the end of the last century had already determined to have its seat in the same parts of the brain which originated the emotions themselves. And these facts were well known to all the early students of phrenology. Dr. Davey, of Bristol, stated to the "Bath and Bristol Medical Association" in 1874, that in 1842 he was present at a series of experiments which went to demonstrate, in the most decided and unequivocal manner, that the stimulation of many parts of the cerebrum of man did excite both sensation and motion. He added:—"I affirm, that twenty-eight years before Hitzig ascertained and taught the fact as stated, the same was known to the late Dr. Elliotson, to the late Dr. Engledue, and to Messrs. Atkinson and Syme, of London, including others who may be nameless. It is not now, as it was then, so really dangerous to announce the discovery of things new and strange. The present age *is*, we hope, less illiberal than I knew and even felt it to be at the time referred to. Doctors Hitzig and Ferrier would not be reaping the happy harvest of their very commendable labours if things were not now altered for the better."

It is clear then that the correspondence of the motor-areas of Ferrier with the phrenological organs of which the particular motions are the natural expression, was discovered by Gall and was well known to all the early phrenologists; but the modern writers, owing to their ignorance of phrenology, have denied this correspondence. It has, however, been clearly pointed out by Mr. James Webb, late President of the British Phrenological Association, in his *Phrenological Aspect of Modern Physiological Research* (1890), and by Dr. Bernard Höllander, M.D., at the British Association in 1890, and before the Anthropological Institute in 1889 and 1891.[1] A few of the examples, beginning with those adduced by Dr. Höllander, will be here summarized, but the original papers must be consulted for the full evidence.

Professor Ferrier excited a definite portion of the ascending frontal convolution in monkeys and several other animals, which had the effect of elevating the cheeks and angles of the mouth with closure of the eyes. On no other region could the same effect be produced. Now the expression of joy or amusement is the drawing back the corners of the mouth forming an incipient smile. All the authorities agree in this. General paralysis of the insane is almost always accompanied by optimism and constant joyousness, accompanied by delusions as to wealth and grandeur; and the earliest physical symptom of the disease is a trembling at the corners of the mouth and the outer corners of the eyes. Now the brain-centre producing these effects corresponds in position to the phrenological organ of Hope, the manifestation of which is cheerfulness and especially cheerful anticipations.

Professor Ferrier also discovered a centre for facial movements, and this exactly corresponds with the

[1] See *Journal of the Anthropological Institute*, vol. xix. p. 12, and vol. xx. p. 227.

phrenological organ of Imitation, which gives the power of mimicry, of which facial expression is the most important part.

Another centre was found which produced motions of the tongue, cheek-pouches and jaws in monkeys, exactly as in tasting; and this spot corresponds with the organ of Gustativeness, which gives appreciations of flavours, and in its excess makes a man a gourmand.

A most remarkable correspondence is that of the organ of Concentrativeness, which gives the power of continued attention to any subject, and is the centre of visual ideation. It is not the centre of *vision*—that is situated in another part of the brain—but of the power of seeing and attending to definite objects. Its outward manifestation is a fixed gaze; and as sight is by far the most important of the senses as regards giving us knowledge of the outer world, concentration of attention would be first developed through vision, and a fixity of gaze has become an outward indication of continuous thought on any subject, even non-visual. The person is said to exhibit "rapt attention."

One more correspondence noted by Dr. Höllander may be given—that of the centre for motions indicating anger with the phrenological organ of Destructiveness. The excitation of this centre caused jackals to retract the ears and spring forward; in cats, opening the mouth, with spitting and lashing the tail—all indications of anger. Now Destructiveness—perhaps badly named—is simply the organ of anger or passion; and unrestrained passion, whether in children or adults, is usually manifested by injury or destruction of the offending object; the child beats or breaks what has hurt it, while the despot tortures or kills the person who seriously angers him.

Mr. Webb gives illustrations from several other organs, which are equally interesting. When Dr. Ferrier's centre (1) was excited in monkeys, the animal "extended its legs." This centre is in the position of the phrenological organs of Firmness and Self-esteem, one outward expression of which is the stretching the legs, or putting down the feet with determination, whence has arisen a proverbial expression for obstinacy. Excitation of centre (12) caused the "eyes to open widely, the pupils to dilate, and head and eyes to turn to the other side." Now this centre corresponds to the phrenological organ of Wonder, and nothing could better express wonder than the motions described. Even more curious was the result of exciting the lower part of the inferior occipital convolution of cows and sheep, which "caused uneasy movements of the hind legs and tail, while the animals looked to the opposite hind leg and occasionally uttered a plaintive cry, as if of pain or annoyance." The part excited is the phrenological organ of Philoprogenitiveness or love of offspring, and any one who has watched a cow whose calf has been taken away, must recognize the accurate description of the motions by which she expresses her feelings.

Now, surely, this close correspondence of "motor-centres" with the phrenological organs of which the actions or motions under excitation are the natural expression, is very remarkable, and affords a new and striking test of the accuracy with which the phrenologists have localised the brain-centres for the various mental faculties. With such confirmation as regards most of the motor-centres yet discovered, the presumption is in favour of the accuracy of the bulk of the phrenological organs, more especially as their development also accords with, and explains, national and race character, which neither physiologists nor anthropologists have even attempted to do; while as regards individual character, the skilled phrenologist has shown that he is able to read it like an open book, and to lay bare the hidden springs of conduct with an accuracy that the most intimate friends of the individual cannot approach. Yet, even now, the advocates of this new and very crude method of brain-study, repeat the old vague objections to phrenology, as if they were true and unanswerable. After the reading of Dr. Höllander's first paper at the Anthropological Institute, Professor Ferrier, while complimenting the author, and making no objections to his facts, went on to say:—"What we wanted was evidence founded on careful investigation according to strictly scientific methods, serving to indicate a relation between the development of particular centres and special mental faculties, aptitudes, or peculiarities. At present he did not think there was any such worthy of consideration." But were not Gall's, and Spurzheim's, and Combe's life-long investigations "careful," and their methods "scientific"? And were not their final conclusions justified by that best test of all true theory, the power of prediction of character in its most minute details? Life-long and class prejudices always die hard, but it is surely now time, that this wholly unjustifiable accusation, of phrenology being "unscientific," should be abandoned, since it is really founded on a far more scientific basis than that of the modern school, who, by an utterly unnatural, and therefore "unscientific" mode of exciting the brains of living animals, hope to arrive at a correct knowledge of its varied functions.

The blinding effects of this prejudice against phrenology, has caused these modern investigators to overlook the circumstance, that the often complex motions of different parts of the body resulting from the stimulation of various brain-centres, were really the physical expression of mental emotions, and of the very same emotions as those long since assigned to the phrenological organs situated in the same parts of the brain. It is also very suggestive that these experiments lead to nothing of value in the hands of the experimenters. To show that the excitation of one brain-centre affects such numerous and varied sets of muscles as are required to cause movements of the hind legs, the tail, the head, and the vocal organs of a cow; while excitation of another centre produces movements of the ear and of all four limbs in a jackal, but of the tail, mouth, and tongue in a not very remote species, the cat—are facts which, standing alone, are unmeaning and worthless. But all these movements and many others become quite intelligible when looked upon as, not the immediate but the derived effects of the stimulation; being the various modes of expression of the mental emotions which constitute the actual functions of the parts excited, and the expression of which varies according to the organization and habits of the several animals. Instead of being, as so often alleged, a disproof of phrenology, or in any way antagonistic to it, these modern investigations are only intelligible when explained by means of its long-established facts, and thus really furnish a most striking and most convincing, because wholly unintended, confirmation of its substantial truth.

Let us now briefly state the main principles of phrenology, all at first denied, but all now forming part of recognized science.

(1) The brain is the organ of the mind.

This was denied in the *Edinburgh Review*, and even J. S. Mill wrote that "mental phenomena do not admit of being deduced from the physiological laws of our nervous organization."

(2) Size is, other things being equal, a measure of power. This was at first denied, but is now generally admitted by physiologists.

(3) The brain is a congeries of organs, each having its appropriate faculty.

Till a comparatively recent period this was denied, and the brain was said to act as a single organ. Now it is admitted that there are such separate organs, but it is alleged that they have not yet been discovered.

(4) The front of the brain is the seat of our perceptive and reflective faculties; the top, of our higher sentiments; the back and sides, of our animal instincts.

This was long denied; even the late Dr. W. B. Carpenter maintained that the back of the brain was probably the seat of the intellect! Now, almost all physiologists admit that this *general* division of brain-organs is correct.

(5) The form of the skull during life corresponds so closely to that of the brain, that it is possible to determine the proportionate development of various parts of the latter by an examination of the former.

The denial of this was, as we have seen, the stock objection to the very possibility of a science of phrenology. Now it is admitted by all anatomists. The late Professor George M. Humphry, of Cambridge University, in his *Treatise on the Human Skeleton*

(p. 207), expressly admits the correspondence, adding—"The arguments against phrenology must be of a deeper kind than this to convince any one who has carefully considered the subject."

It thus appears that the five main contentions of the phrenologists, each of them at first strenuously denied, have now received the assent of the most advanced modern physiologists. But admitting these fundamental data, it evidently becomes a question solely of a sufficiently extended series of comparisons of *form* with *faculty* to determine what faculties are constantly associated with a superior development of any portion of the cranium and of the brain within it. To assert that such comparisons are unscientific, without giving solid reasons for the assertion, is absurd. The whole question is, are they adequate? And the one test of adequacy is, do they enable the well-instructed student to determine the character of individuals from the form of their skulls, whenever any organ or group of organs are much above or below the average? This test was applied to the early phrenologists in scores, in hundreds, even in thousands of cases, with a marvellous proportion of successful results. The men who first determined the position of each organ, only did so after years of observation and hundreds of comparisons of development of organ with manifestation of function. These determinations were *never* blindly accepted, but were tested by their followers in every possible way, and were only generally admitted when every ordeal had been passed successfully. To reject such determinations without full examination of the evidence in support of them, without applying any of the careful tests which the early phrenologists applied, and on the mere vague allegations of insufficient observation or unscientific method, is itself utterly unscientific.

In the coming century Phrenology will assuredly attain general acceptance. It will prove itself to be the true science of mind. Its practical uses in education, in self-discipline, in the reformatory treatment of criminals, and in the remedial treatment of the insane, will give it one of the highest places in the hierarchy of the sciences; and its persistent neglect and obloquy during the last sixty years, will be referred to as an example of the almost incredible narrowness and prejudice which prevailed among men of science, at the very time they were making such splendid advances in other fields of thought and discovery.

# CHAPTER XVII

## THE OPPOSITION TO HYPNOTISM AND PSYCHICAL RESEARCH

Speak gently of the new-born gift, restrain the scoff and sneer,
And think how much we may not learn is yet around us here;
What paths there are where faith must lead, and knowledge cannot share,
Though still we tread the devious way, and feel that truth is there.

—*Anon.* (1844).

Sleep, sleep on! forget thy pain;
My hand is on thy brow,
My spirit on thy brain;
My pity on thy heart, poor friend;
And from my fingers flow
The powers of life, and like a sign,
Seal thee from thine hour of woe.

—*Shelley.*

ALTHOUGH the subjects to be now discussed have made some progress in the last quarter of the century, this was preceded by a long period of ignorance, accompanied by the most violent opposition, extremely discreditable to an age of such general research and freedom of inquiry in all other branches of human knowledge. A brief outline of the nature of this opposition will be interesting; and may serve as a warning to those who still put faith in the denunciations of the public press, or of those writers who pose as authorities without having devoted any serious study to the subject.

The phenomena of Animal Magnetism, often termed Mesmerism, and now Hypnotism, were discovered by a physician of Vienna named Mesmer about the year 1770. He applied it to the treatment of disease, and obtained great popularity in Paris, where he came to practise. His knowledge of the subject was, however, necessarily limited, and his interpretation of the facts often erroneous. A Government Commission was appointed in 1785, consisting of physicians and scientists (including Lavoisier, Franklin, and other eminent men) who, finding that many of the phenomena, alleged by Mesmer to be due to a special form of magnetism, could be produced in the patients by suggestion, reported against his alleged powers, and the subject soon fell into disrepute.

Early in the present century, however, the phenomena again occurred in the practice of some physicians in Paris and elsewhere, a few of whom devoted much time to the study, and obtained evidence of the most perfect thought-reading, true clairvoyance, and many other apparently superhuman powers. Many medical men became satisfied of the genuineness of these strange occurrences, and the amount of interest they excited in the scientific and medical worlds is shown by the fact, that the article "Magnetisme" in the *Dictionnaire de Médecine*, published in 1825, treated the subject in a serious spirit, and recognised the whole of its phenomena as being undoubtedly genuine. The writer, Dr. Rostan, declares that he had himself examined a clairvoyante who, when he placed his watch at the back of her head, told the time indicated by it, and even when he turned the hands round without looking at them, was equally successful.

Of course those who had no opportunity of investigating the subject under favourable conditions, could not accept such marvels, and imputed them to clever trickery; and in order to determine authoritatively how much truth there was in the statements of the Animal Magnetisers, the Académie Royale de Médecine, in 1826, appointed a committee of eleven members, all, of course, medical men, and presumably capable and impartial, to inquire into the whole subject experimentally. Nine of the members attended the meetings and experiments during five years; and in 1831 they delivered a full and elaborate Report, which was signed by the whole nine, and was therefore unanimous. This Report (published in the *Archives Generale de Médecine*, vol. xx.) gives the details of a large number of experiments, and concludes with a summary of what was considered to be proved, together with some weighty observations. As this Report is very little known, and has been completely ignored by almost all writers adverse to the claims of the magnetisers, I will give some of the more important portions of it, as translated by Dr. Lee in his work on Animal Magnetism.

### *Report of the Commission of the Académie Royale de Médecine on Animal Magnetism.*

"Conclusions and General Remarks."

"The commission has reported with impartiality that which it had seen with distrust; it has exposed methodically that which it has observed under different circumstances, and which it has followed up with an attention as close as it is continued. It has the consciousness that the statements which it presents to you are the faithful expression of that which it has observed. The obstacles which it has met with are known to you; they are partly the cause of the delay which has occurred in presenting the report, although we have long been in possession of the materials. We are, however, far from excusing ourselves, or from complaining of this delay, since it gives to our observations a character of maturity and reserve which should lead you to confide in the facts which we have related, without the charge of prepossession and enthusiasm with which you might have reproached us if we had only recently collected them. We add that we are far from thinking that we have seen all that is to be seen, and we do not pretend to lead you to admit as an axiom, that there is nothing positive in magnetism beyond what we mention in our report. Far from placing limits to this part of physiological science, we entertain, on the contrary, the hope that *a new field* is opened to it; and guaranteeing our own observations, presenting them with confidence to those who, after us, will occupy themselves with magnetism, we restrict ourselves to drawing the following conclusions, which are the necessary consequence of the facts the totality of which constitutes our report."

A considerable proportion of these "conclusions" relate to points which are either unimportant or now undisputed, such as the mode of magnetising, the proportion of persons who can be magnetised, the influence of expectation, the variety of the phenomena produced, the possibility of simulation, the nature of the magnetic sleep, the therapeutic effects produced and their importance, and other similar points. The following paragraphs give the more important of the "conclusions" referring to those points which are still doubted or denied by a considerable number of men of science.

"It has been demonstrated to us that the magnetic sleep may be produced under circumstances in which the magnetised have not been able to perceive, and have been ignorant of, the means employed to occasion it."

"When a person has been already magnetised, it is not always necessary to have recourse to contact, or to the 'passes,' in order to magnetise afresh. The look of the magnetiser, his will alone, has often the same influence. In this case one can not only act upon the magnetised, but throw him completely into the sleep, and awaken him from this state without his being aware of it, out of his sight, at a certain distance, and through closed doors."

* * * * *

"We have seen two somnambulists distinguish *with closed eyes* the objects placed before them; they have designated, *without touching them*, the colour and name of cards; they have read words written, or lines from a book. This phenomenon has occurred even when the eyelids *were kept closed by the fingers.*"

"We have met with two somnambulists who possessed the faculty of foreseeing acts of the organism, more or less distinct, more or less complicated."

"We have only met with one somnambulist who

could indicate the symptoms of the diseases of three persons with whom she was placed in relation. We had, however, made researches on a considerable number."

* * * * *

"The commission could not verify, because it had no opportunity, the other faculties which magnetisers had stated to exist in somnambulists. But it has collected, and it communicates to the Academy, facts sufficiently important to induce it to think that the Academy ought to encourage researches on magnetism as a very curious branch of psychology and natural history.

"Certainly we dare not flatter ourselves that we shall make you share entirely our conviction of the reality of the phenomena which we have observed, and which you have neither seen, nor followed, nor studied with or in opposition to us. We do not therefore exact from you a blind belief in all that we have reported. We conceive that a great part of the facts are so extraordinary that you cannot grant it to us: perhaps we ourselves should have refused you our belief, if, changing places, you had come to announce them before this tribunal to us, who, like you at present, had seen nothing, observed nothing, studied nothing, followed nothing of them.

"We only require you to judge us as we should have judged you, that is to say, that you remain perfectly convinced that neither the love of the wonderful, nor the desire of celebrity, nor any interest whatever, has influenced our labours. We were animated by motives more elevated, more worthy of you—by the love of science and by the wish to justify the hopes which the Academy had conceived of our zeal and devotedness."

"(Signed) BOURDOIS DE LA MOTTE (*President*),
FOUQUIER,
GUENEAU DE MUSSY,
GUERSENT,
ITARD,
LEROUX,
MARC,
THILLAGE,
HUSSON (*Reporter*)."

It is hardly possible to have a weightier or more trustworthy report than this one, showing in every line the care and deliberation of the members of the commision, while their competence and honesty are above suspicion. The same general conclusions as to the reality and importance of animal magnetism were arrived at by some of the most eminent physicians in Russia, Denmark, Saxony and other countries; while the entire report of the French Commission was translated into English in 1836, and published in Mr. Colquhoun's *Isis Revelata.*

In 1837, however, in consequence of many accounts of clairvoyance then occurring in various parts of France, the Académie de Médecine offered a prize of three thousand francs to any one who should prove his ability to read without use of the eyes. The daughter of a physician at Montpelier—Dr. Pigeaire—possessed this power, as testified by many persons of repute; and, in consequence of this offer, he brought her to Paris. Many persons saw her in private, and several physicians—MM. Orfila, Ribes, Reveillè-

Parisé and others—certified the fact of her clairvoyant powers. But the members appointed by the Academy —less experienced than those of the Commission of 1831—began by making stipulations as to the complete enclosure of the clairvoyant's head, to which her father would not consent, and thus the opportunity of officially testing this lady was lost.[1] Others presented themselves, but none succeeded. The result was therefore purely negative; but as there were in some cases suspicions of imposture or attempts at imposture, the report was, of course, against the existence of clairvoyance. This was only what might have been anticipated by all who had really investigated the subject. Professor William Gregory, of the University of Edinburgh, after twenty years' study of animal magnetism and an extensive personal experience, wrote as follows:—

"In regard to clairvoyance, I have never seen it satisfactorily exhibited except quite in private; and in this point my experience has simply confirmed the statements made by the best observers. I feel confident that every one who chooses to devote some time and labour to the investigation, may meet with it, either in his own cases or those of his friends."

In his *Letters on Animal Magnetism* Professor Gregory gives several indisputable cases tested by himself. Dr. Haddock, Major Buckley, Sir Walter Trevelyan, Miss Martineau, Dr. Esdaile, Dr. Lee, and Dr. Elliotson, have all obtained evidence of the most convincing

[1] The method usually adopted was to bind a linen cloth over the eyes, to cover this with cotton-wool, and over all a black velvet mask, which was held to be a complete test by Arago and other observers. This, however, the commissioners would not even try.

kind, much of which has been published; while many eminent physicians and men of science on the Continent obtained equally convincing results—all confirming the positive evidence of the French Commission of 1831, and proving that the negative results of the Commission of 1837 were due to the inexperience and prejudices of the members. Yet, notwithstanding this cumulative proof, modern writers against the higher phenomena produced by hypnotism, appear to be either totally ignorant of the existence of the five years' inquiry and elaborate report of the first commission of the Académie de Médecine, or confound it with the second commission, which gave a purely negative report on one limited phase of the phenomena!

Thus, the late Dr. W. B. Carpenter, in his volume on *Mesmerism, Spiritualism, etc., Historically and Scientifically Considered* (Longmans, 1877), writes as follows:—

"It was in France that the pretensions of mesmeric *clairvoyance* were first advanced; and it was by the French Academy of Medicine, in which the mesmeric state had been previously discussed with reference to the performance of surgical operations, that this new and more extraordinary claim was first carefully sifted, in consequence of the offer made in 1837 by M. Burdin (himself a member of that Academy) of a prize of 3,000 francs to any one who should be found capable of reading through opaque substances."

Neither here, nor in any part of his volume, does Dr. Carpenter show any knowledge of the existence of the Commission of 1825–31, which really "first carefully sifted" the varied phenomena of Animal Magnetism, including numerous cases of clairvoyance, and decided that they were genuine.

In the last edition of Chambers' *Encyclopedia*, a publication remarkable for the great ability of its contributors and the impartial treatment of disputed questions, we find in the article "Animal Magnetism" the following passage:—"Despite the unfavourable report of the French Commission of 1785, as well as of a later one in 1831, and other subsequent exposures" . . . —indicating that the writer was unacquainted with the favourable report of 1831, and confused it with the negative report of 1837–40. And this ignorance is confirmed by the statement, a little further on, that "no scientific observer has yet confirmed the statements of mesmerists as to clairvoyance, reading of sealed letters, influence on unconscious persons at a distance, or the like,"—a statement the exact opposite of the fact, since the nine members of the commission of the Academy of Medicine, Professor Gregory and the other gentlemen mentioned above, as well as a large number of physicians and others on the Continent, must surely be held to be, individually and as a whole, "scientific observers," or the term can have no meaning. Büchner, Spitta, and other antagonistic continental writers, also appealed to the commission of 1784 as having exposed "the swindle of magnetic cures," apparently in complete ignorance of the report of 1831; and Büchner also refers to the commission of 1837 as reporting against clairvoyance, without any reference to the more weighty report of 1831 in its favour.

One more example as to the mode of treatment of evidence for the reality of clairvoyance. Dr. Carpenter describes some of his own visits to Alexis and Adolphe Didier, accompanied by Dr. Forbes; and because they saw nothing which was to them absolutely conclusive, he leads the reader to think that nothing really conclusive had ever been obtained. But Dr. Lee, a physician of repute, and therefore presumably as good a witness as Dr. Carpenter or Dr. Forbes, in his well-known work on Animal Magnetism, devotes twenty-two pages to an account of his own personal experiments with Alexis at Brighton in 1849, including such a number and variety of striking tests as to entirely outweigh any number of *negative* results like those of Dr. Carpenter. And in addition to these, other special tests of the most stringent character have been published, two of which may be here given. Sergeant Cox, in his *What Am I?* (vol. ii. p. 167) describes a test by a party of experts of whom he was one. A word was written by a friend in a distant town, and enclosed in an envelope, *without any one of the party knowing what the word was.* This envelope was enclosed successively in six others of thick brown paper, each sealed. This packet was handed to Alexis, who placed it on his forehead, and in three minutes and a half wrote the contents correctly, imitating the very handwriting. Let any one compare Dr. Carpenter's explanation of how he supposed such readings were done, and he will see how completely inadequate it is as applying to tests such as that of Sergeant Cox and scores of other inquirers.

The next test is furnished by the experience of the greatest of modern professional conjurers, Houdin, who, at the request of the Marquis de Mirville, had two sittings with Alexis. His account, as quoted by Dr. Lee, is as follows. After describing what took place at the first sitting, he says: "I cannot help declaring that the facts here reported are perfectly exact, and that the more I reflect upon them, the more impossible do I find it to class them with those which constitute the object of my art." (May 10th, 1849.)

"At the second *seance* I witnessed still more surprising events than at the first, and they no longer leave any doubt in my mind respecting the clairvoyance of Alexis. I tear off the envelope of a pack of cards I brought with me. I shuffle and deal with every precaution, which, however, is useless, for Alexis stopped me by naming a card which I had just placed before him on the table. 'I have the king,' said he. 'But you know nothing about it, as the trump card is not turned up.' 'You will see,' he replied; 'go on.' In fact, I turned up the ace of spades, and his card was the king of spades. The game was continued; he told me the cards which I should play, though my cards *were held closely in my hands beneath the table.* To each of the cards I played he followed suit, *without turning up his cards*, which were always perfectly in accordance with those I led. I therefore returned from this *séance* as astonished as one can be, and I am convinced that it is quite impossible that chance, or any superior skill, could produce such wonderful results." (May 16th, 1849.)

Now the point which I wish to submit to my readers is, whether the method of argument and discussion adopted by the most eminent opponents of Animal Magnetism, is either honest, or scientific, or even rational. We do not ask them to accept blindly any of the facts reported, or to refrain from any criticism, however severe, which is founded upon a fair consideration of all the available evidence. But in this matter, as I have here shown by a few striking examples, the public mind is influenced by the omission to state the case fairly; by putting forth the weakest instead of the strongest facts and arguments; and by the denial that any good and trustworthy evidence exists. What should we think of the man who discussed any of the disputed questions of recognised science in this way? who either ignorantly or wilfully omitted all reference to the most careful researches of the most eminent writers on the subject; and while professing to instruct and enlighten the public, led them to believe that such researches did not exist? Such a man would at once lose all claim to be considered an authority on any subject, and his future writings would be treated with deserved neglect. It is because, during the greater part of the century, this most important and most interesting inquiry has been treated in so unworthy a manner by men of reputation in other departments of research, that we are compelled to class the opposition to the phenomena of mesmerism, and especially to the reality of clairvoyance, as constituting one of the exceptions to the steady march of most branches of science throughout the century.

We now come to the consideration of a practical application of animal magnetism, the opposition to which was even more virulent and more unjustifiable than that just described. The subject of Mesmerism, as it began to be termed, was first introduced into this country by Mr. Richard Chenevix, a Fellow of the Royal Society, who published a series of papers in the *London Medical and Physical Journal* in 1829. He also exhibited the phenomena to numerous medical men, among others to Dr. Elliotson, who afterwards became one of the chief teachers of the science. The Professor of Physiology at King's College, Dr. Mayo, also upheld and wrote upon it in the medical journals. Baron Dupotet came to London and again demonstrated the main facts, as did numbers of public-lecturers, affording ample opportunities for experiment and observation.

In 1829, M. Cloquet, one of the most eminent surgeons of Paris, amputated a cancerous breast during the mesmeric sleep, the patient being entirely insensible to pain, although able to converse. Teeth were extracted, and many other operations, some very serious, such as the extirpation of a portion of the lower jaw in the hospital of Cherbourg, were performed in France. About twelve years later, operations in the mesmeric trance began to be performed in England; but, notwithstanding the numerous cases already reported from France, supporting the fact of insensibility to pain, as fully described by the Academy of Medicine, they were received with general incredulity by the medical profession, while the most outrageous accusations were made against all who took part in them.

On the 22nd of November, 1842, at the Royal Medical and Chirurgical Society of London, an account was read of the amputation of the thigh during the mesmeric trance. The patient was a labourer who had suffered for five years with neglected disease of the left knee, the slightest motion of the joint being attended with extreme pain. Before the operation he had had no sleep for three nights. He was mesmerised by Mr. W. Topham, a barrister, and operated upon by Mr. W. Squire Ward, surgeon, in the District Hospital of Wellow, Nottinghamshire. During the whole operation, lasting twenty minutes, the patient remained in perfect repose, the placid countenance never changing, while no muscle of the body or limbs was seen to twitch. He awoke gradually and calmly, and on being questioned, declared that he knew nothing that was being done, and had felt no pain at all. He recovered perfectly, and had not a single bad symptom.

Then followed a violent discussion. Mr. Coulson said the non-expression of pain was a common thing, and he had no doubt the man had been trained to it. Several declared that the man shammed. One declared he would not have believed the facts had he witnessed them! Then the great men of the profession spoke. Dr. Marshall Hall, the investigator of reflex-action, declared that it was a case of imposition, because the sound leg should have contracted when the diseased leg was cut. The case, therefore, contradicted itself. Sir Benjamin Brodie believed that the man must have been naturally insusceptible of pain. He also agreed with Dr. Marshall Hall that the other leg ought to have moved, and he was quite satisfied with the two French reports against mesmerism. Mr. Liston and Mr. Bransby Cooper made fun of the subject; but Dr. Mayo declared it was a paper of great importance, and should not be ridiculed. Mr. Wood, who had assisted at the amputation, vouched for the complete accuracy of the whole account, and pointed out that before the operation the patient *had* suffered intense pain, and that during the operation he not only showed no sign of pain, but no sign of resistance to the expression of pain. Dr. Elliotson also pointed out the illogical nature of the objections; but the opponents, who were all completely ignorant of the subject, at the next meeting refused confirmation of the minutes, which were therefore expunged!

Here we have extreme ignorance in high places, denying facts which had been observed again and again by men as honest and trustworthy as themselves. It was these men, and others equally ignorant, who accused the operators of bribing their patients not to exhibit pain; who accused Dr. Elliotson of "polluting the temple of science"; and who ejected this eminent physician from his professorship in the University of London, because he persisted in studying the phenomena of mesmerism and in publishing the results of his experiments. He was, however, soon justified in the eyes of all the more honest members of the profession by the publication of so many cases of painless operations as to compel their acceptance as facts;[1] while he was supported by Dr. Esdaile, who gave an account of more than 300 operations performed by himself and other surgeons in the hospitals of Calcutta, which were confirmed by a commission appointed to inquire into them by the Bengal Government, and by the Governor-General himself. The reports of these cases showed that the patients were equally subject to the charge of imposition because they did not exhibit reflex-action in the opposite limb; and Dr. Elliotson made this point the subject of some justifiable ridicule. He says: "It is really lamentable to know that this Asiatic practised imposition as boldly as the female in Europe. The Indian was convicted through the self-same piece of ignorance. He too was unaware that he ought to have moved his right elbow-joint, if he felt nothing while his left was being cut off;

[1] *Numerous Cases for Surgical Operations without Pain in the Mesmeric State*, by John Elliotson, M.D., F.R.S., London, 1843.

and so he did not stir it. The dark races are just as wicked and just as ignorant of physiology as the white."

The facts, however, accumulated so rapidly and were so well attested, that a few years later Dr. Noble, Sir John Forbes, and Dr. W. B. Carpenter accepted them; thus admitting that the great men who denied them were wholly in the wrong, and that they had displayed ignorance and prejudice in their accusations of imposture and bad faith. But just when the great importance of mesmerism in rendering the most serious operations painless, and at the same time greatly assisting the patient's recovery, was fully acknowledged, the discovery of anæsthetics occurred; and this physiological agent, being more easy to apply and more certain to act upon all patients, soon led to the neglect of mesmerism. With this neglect the old prejudices and incredulity revived; and, although its soothing and remedial influence in disease was quite as well established as its use in surgery, it soon fell into disuse, and the great majority of medical men came to look upon it as either disreputable or altogether a delusion. For nearly half a century it remained it abeyance, till its study was revived in the French hospitals, where all the phenomena described by the early mesmerisers have been re-observed, together with some others even more extraordinary.

During the latter portion of the century, the study of these and other obscure psychical phenomena has become more extended, and in every civilized country societies have been formed for investigation, and many remarkable works have been published. One after another, facts, long denied as delusions or exaggerations, have been admitted to be realities. The stigmata, which at different times have occurred in Catholic countries, are no longer sneered at as priestly impostures. Thought-transference, automatic-writing, trance-speaking, and clairvoyance, have been all demonstrated in the presence of living observers of undoubted ability and knowledge, as they were demonstrated to the observers of the early part of the century and carefully recorded by them. The still more extraordinary phenomena—veridical hallucinations, warnings, detailed predictions of future events, phantoms, voices or knockings, visible or audible to numerous individuals, bell-ringing, the playing on musical instruments, stone-throwing, and various movements of solid bodies, all without human contact or any discoverable physical cause—still occur among us as they have occurred in all ages. These are now being investigated, and slowly but surely are proved to be realities, although the majority of scientific men and of writers for the press still ignore the cumulative evidence, and ridicule the inquirers. These phenomena, being comparatively rare, are as yet known to but a limited number of persons; but the evidence for their reality is already very extensive, and it is absolutely certain that, during the coming century, they too will be accepted as realities by all impartial students and by the majority of educated men and women.

The great lesson to be learnt from our review of this subject is, distrust of all *à priori* judgments as to *facts*; for the whole history of the progress of human knowledge, and especially of that department of knowledge now known as psychical research, renders it certain, that, whenever the scientific men or popular teachers of any age have denied, on *à priori* grounds of impossibility or opposition to the "laws of nature," the facts observed and recorded by numerous investigators of average honesty and intelligence, these deniers *have always been wrong.*[1]

Future ages will, I believe, be astonished at the vast amount of energy and ignorance displayed by so many of the great men of this century in opposing unpalatable truths, and in supposing that *à priori* arguments, accusations of imposture or insanity, or personal abuse, were the proper means of determining matters of fact and of observation in any department of human knowledge.

[1] For a discussion of this point with illustrative cases see my *Miracles and Modern Spiritualism*, pp. 17–29.

# CHAPTER XVIII

## VACCINATION A DELUSION—ITS PENAL ENFORCEMENT A CRIME

To-day in all its dimpled bloom,
  The rosy darling crows with glee;
To-morrow in a darkened room
  A pallid, wailing infant see,
Whose every vein from head to heel
Ferments with poison from my steel.
—*A. H. Hume.*

Against the body of a healthy man, Parliament has no right of assault whatever, under pretence of the Public Health; nor any the more against the body of a healthy infant.
—*Prof. F. W. Newman.*

## I

### VACCINATION AND SMALL-POX

Among the greatest self-created scourges of civilized humanity are the group of zymotic diseases, or those which arise from infection, and are believed to be due to the agency of minute organisms which rapidly increase in bodies offering favourable conditions, and often cause death. Such diseases are: plague, small-pox, measles, whooping-cough, yellow fever, typhus and enteric fevers, scarlet fever and diphtheria, and cholera. The conditions which especially favour these diseases are foul air and water, decaying organic matter, overcrowding, and other unwholesome surroundings whence they have been termed "filth diseases." The most terrible and fatal of these—the plague—prevails only where people live under the very worst sanitary conditions as regards ventilation, water supply, and general cleanliness. Till about 250 years ago it was as common in England as small-pox has been during the present century, but a very partial and limited advance in healthy conditions of life entirely abolished it, its place being to some extent taken by small-pox, cholera, and fevers. The exact mode by which all these diseases spread is not known; cholera, diphtheria, and enteric fever are believed to be communicated through the dejecta from the patient contaminating drinking water. The other diseases are spread either by bodily contact or by transmission of germs through the air; but with all of them there must be conditions favouring their reception and increase. Not only are many persons apparently insusceptible through life to some of these diseases, but all the evidence goes to show that, if the whole population of a country lived under thoroughly healthy conditions as regards pure air, pure water, and wholesome food, none of them could ever obtain a footing, and they would die out as completely as the plague and leprosy have died out, though both were once so prevalent in England.

But during the last century there was no such knowledge, and no general belief in the efficacy of simple, healthy conditions of life as the only effectual safeguard against these diseases. Small-pox, although then, as now, an epidemic disease and of very varying degrees of virulence, was much dreaded, because, owing chiefly to improper treatment, it was often fatal, and still more often produced disfigurement or even blindness. When, therefore, the method of inoculation was introduced from the East in the early part of the eighteenth century, it was quickly welcomed, because a mild form of the disease was produced which rarely caused death or disfigurement, though it was believed to be an effectual protection against taking the disease by ordinary infection. It was, however, soon found that the mild small-pox usually produced by inoculation was quite as infectious as the natural disease, and became quite as fatal to persons who caught it. Towards the end of the last century many medical men became so impressed with its danger that they advocated more attention to sanitation and the isolation of patients, because inoculation, though it may have saved individuals, really increased the total deaths from small-pox.

Under these circumstances we can well understand the favourable reception given to an operation which produced a slight, *non-infectious* disease, which yet was alleged to protect against small-pox as completely as did the inoculated disease itself. This was Vaccination, which arose from the belief of farmers in Gloucestershire and elsewhere that those who had caught cow-pox from cows were free from small-pox for the rest of their lives. Jenner, in 1798, published his *Inquiry*, giving an account of the facts which, in his opinion, proved this to be the case. But in the light of our present knowledge we see that they are wholly inconclusive. Six of his patients had had cow-pox when young, and were inoculated with small-pox in the usual way from twenty-one to fifty-three years afterwards, and because they did not take the disease, he concluded that the cow-pox had preserved them. But we know that a considerable proportion of persons in middle age are insusceptible to small-pox infection; besides which, even those who most strongly uphold vaccination now admit that its effects die out entirely in a few years—some say four or five, some ten—so that these people who had had cow-pox so long before were certainly *not* protected by it from taking small-pox. Several other patients were farriers or stable men who were infected by horse-grease, not by cow-pox, and were also said to be insusceptible to small-pox inoculation, though not so completely as those who had had cow-pox. The remainder of Jenner's cases were six children, from five to eight years old, who were vaccinated, and then inoculated a few weeks or months afterwards. These cases are fallacious from two causes. In the first place, any remnant of the effects of the vaccination (which were sometimes severe), or the existence of scurvy, then very prevalent, or of any other skin-disease, might prevent the test-inoculation from producing any effect.[1] The other

[1] Professor Crookshank, in his evidence before the Royal Commission (4th Report, Q. 11,729) quotes Dr. De Haën, a writer on Inoculation, as saying: "Asthma, consumption, hectic or slow fever of any kind, internal ulcers, obstructed glands, obstructions of the viscera from fevers, scrofula, scurvy, itch, eruptions, local inflammations or pains of any kind, debility, suppressed or irregular menstruation, chlorosis, jaundice, pregnancy, *lues venerea*, whether in the parent or transmitted to the child, and a constitution under the strong influence of mercury, prevented the operation." There is no evidence that those who applied the so-called "variolous test" in the early days of vaccination paid any attention to this long list of ailments, many of which were very prevalent at the time, and which would, in the opinion of De Haën, and of the English writer Sanders, who quotes him, have prevented the action of the virus and thus rendered the "test" entirely fallacious. With such causes as these, added to those already discussed, it becomes less difficult to understand how it was that the alleged test was thought to prove the influence of the previous vaccination without really doing so.

cause of uncertainty arises from the fact that this "variolous test" consisted in inoculating with small-pox virus obtained from the *last* of a series of successive patients in whom the effect produced was a minimum, consisting of very few pustules, sometimes only one, and a very slight amount of fever. The results of this test, whether on a person who had had cow-pox or who had not had it, was usually so slight that it could easily be described by a believer in the influence of the one disease on the other as having "no effect"; and Dr. Creighton declares, after a study of the whole literature of the subject, that the description of the results of the test is almost always loose and general, and that in the few cases where more detail is given the symptoms described are almost the same in the vaccinated as in the unvaccinated. Again, no careful tests were ever made by inoculating at the same time, and in exactly the same way, two groups of persons of similar age, constitution, and health, the one group having been vaccinated the other not, and none of them having had small-pox, and then having the resulting effects carefully described and compared by independent experts. Such "control" experiments would now be required in any case of such importance as this; but it was never done in the early days of vaccination, and it appears never to have been done to this day. The alleged "test" was, it is true, applied in a great number of cases by the early observers, especially by Dr. Woodville, physician to a small-pox hospital; but Dr. Creighton shows reason for believing that the lymph he used was contaminated with small-pox, and that the supposed vaccinations were really inoculations. This lymph was widely spread all over the country, and was supplied to Jenner himself, and we thus have explained the effect of the "vaccination" in preventing the subsequent "inoculation" from producing much effect, since both were really mild forms of small-pox inoculation. This matter is fully explained by Dr. Creighton in his evidence before the Royal Commission, printed in the Second Report. Professor E. M. Crookshank, who has made a special study of cow-pox and other animal diseases and their relation to human small-pox, gives important confirmatory evidence, to be found in the Fourth Report.

This brief statement of the early history of vaccination has been introduced here in order to give what seems to be a probable explanation of the remarkable fact, that a large portion of the medical profession accepted, as proved, that vaccination protected against a subsequent inoculation of small-pox, when in reality there was no such proof, as the subsequent history of small-pox epidemics has shown. The medical and other members of the Royal Commission could not realize the possibility of such a failure to get at the truth. Again and again they asked the witnesses above referred to to explain how it was possible that so many educated specialists could be thus deceived. They overlooked the fact that a century ago was, as regards the majority of the medical profession, a prescientific age; and nothing proves this more clearly than the absence of any systematic "control" experiments, and the extreme haste with which some of the heads of the profession expressed their belief in the lifelong protection against small-pox afforded by vaccination, only four years after the discovery had been first announced. This testimony caused Parliament to vote Jenner £10,000 in 1802.

Ample proof now exists of the fallacy of this belief, since vaccination gives no protection (except perhaps for a month or two) as will be shown later on. But there was also no lack of proof in the first ten years of the century; and had it not been for the unscientific haste of the medical witnesses to declare that vaccination protected against small-pox during a whole lifetime—a fact of which they had not and could not possibly have any evidence—this proof of failure would have convinced them and have prevented what is really one of the scandals of the nineteenth century. These early proofs of failure will be now briefly indicated.

Only six years after the announcement of vaccination, in 1804, Dr. B. Moseley, Physician to Chelsea Hospital, published a small book on the cow-pox, containing many cases of persons who had been properly vaccinated and had afterwards had small-pox; and other cases of severe illness, injury, and even death resulting from vaccination; and these failures were admitted by the Royal Jennerian Society in their Report in 1806. Dr. William Rowley, Physician to the St. Marylebone Infirmary, in a work on *Cow-pox Inoculation* in 1805, which reached a third edition in 1806, gave particulars of 504 cases of small-pox and injury after vaccination, with seventy-five deaths. He says to his brother medical men: "Come and see. I have lately had some of the worst species of malignant small-pox in the Marylebone Infirmary, which many of the faculty have examined and know to have been vaccinated." For two days he had an exhibition in his Lecture Room of a number of children suffering from terrible eruptions and other diseases after vaccination.

Dr. Squirrel, formerly Resident Apothecary to the Small-pox and Inoculation Hospital, also published in 1805 numerous cases of small-pox, injuries, and death after vaccination.

John Birch, a London surgeon, at first adopted vaccination and corresponded with Jenner, but soon, finding that it did not protect from small-pox and that it also produced serious and sometimes fatal diseases, he became one of its strongest opponents, and published many letters and pamphlets against it up to the time of his death in 1815.

Mr. William Goldson, a surgeon at Portsea, published a pamphlet in 1804, giving many cases in his own experience of small-pox following vaccination. What made his testimony more important was that he was a believer in vaccination, and sent accounts of some of his cases to Jenner so early as 1802, but no notice was taken of them.[1]

Mr. Thomas Brown, a surgeon of Musselburgh, published in 1809 a volume giving his experiences of the results of vaccination. He had at first accepted and practised it. He also applied the "variolous test" with apparent success, and thereafter went on vaccinating in full confidence that it was protective against small-pox, till 1808, when, during an epidemic, many of his patients caught the disease from two to eight years after vaccination. He gives the details of forty-eight cases, all within his own personal knowledge, and he says he knew of many others. He then again tried the "variolous test," and found twelve cases in which it entirely failed, the result being exactly as

[1] The cases of failure of vaccination here referred to are given in Mr. William White's *Story of a Great Delusion*, where fuller extracts and references will be found.

with those who were inoculated without previous vaccination. These cases, with extracts from Brown's work, were brought before the Royal Commission by Professor Crookshank. (See 4th Report, Q. 11,852.)

Again, Mr. William Tebb brought before the Commission a paper by Dr. Maclean, in the *Medical Observer* of 1810, giving **535** cases of small-pox after vaccination, of which **97** were fatal. He also gave **150** cases of diseases from cow-pox, with the names of ten medical men, including two Professors of Anatomy, who had suffered in their own families from vaccination. The following striking passage is quoted:—"*Doctrine.*—Vaccination or Cow-pox inoculàtion is a perfect preventive of small-pox during life. (Jenner, etc.) *Refutation.*—535 cases of small-pox after cow-pox. *Doctrine.*—Cow-pox renders small-pox milder. It is never fatal. *Refutation.*—97 deaths from small-pox after cow-pox and from cow-pox diseases."

The cases here referred to, of failure of vaccination to protect even for a few years, are probably only a small fraction of those that occurred, since only in exceptional cases would a doctor be able to keep his patients in view, and only one doctor here and there would publish his observations. The controversy was carried on with unusual virulence, hence perhaps the reason why the public paid so little attention to it. But unfortunately both the heads of the medical profession and the legislature had committed themselves by recognising the full claims of Jenner at too early a date and in a manner that admitted of no recall. In 1802, as already stated, the House of Commons, on the Report of its Committee, and the evidence of the leading physicians and surgeons of London—a large number of whom declared their belief that cow-pox was a perfect security against small-pox—voted Jenner £10,000. When therefore the flood of evidence poured in, showing that it did *not* protect, it was already too late to remedy the mischief that had been done, since the profession would not so soon acknowledge its mistake, nor would the legislature admit having hastily voted away the public money without adequate reason. The vaccinators went on vaccinating, the House of Commons gave Jenner £20,000 more in 1807, endowed vaccination with £3,000 a year in 1808, and after providing for free vaccination in 1840, made the operation compulsory in 1853 by a fine, and ordered the Guardians to prosecute in 1867.

### *Vaccination and the Medical Profession*

Before proceeding to adduce the conclusive evidence that now exists of the failure of vaccination, a few preliminary misconceptions must be dealt with. One of these is, that as vaccination is a surgical operation to guard against a special disease, medical men can alone judge of its value. But the fact is the very reverse, for several reasons. In the first place, they are interested parties, not merely in a pecuniary sense, but as affecting the prestige of the whole profession. In no other case should we allow interested persons to decide an important matter. Whether iron ships are safer than wooden ones is not decided by ironmasters or by shipbuilders, but by the experience of sailors and by the statistics of loss. In the administration of medicine or any other remedy for a disease, the conditions are different. The doctor applies the remedy and watches the result, and if he has a large practice he thereby obtains knowledge and experience which no other persons possess. But in the case of vaccination, and especially in the case of public vaccinators, the doctor does not see the result except by accident. Those who get small-pox go to the hospitals, or are treated by other medical men, or may have left the district; and the relation between the vaccination and the attack of small-pox can only be discovered by the accurate registration of all the cases and deaths, with the facts as to vaccination or revaccination. When these facts are accurately registered, to determine what they teach is not the business of a doctor but of a statistician, and there is much evidence to show that doctors are bad statisticians, and have a special faculty for misstating figures. This allegation is so grave and so fundamental to the question at issue, that a few facts must be given in support of it.

The National Vaccine Establishment, supported by Government grants, issued periodical Reports, which were printed by order of the House of Commons; and in successive years we find the following statements:

In 1812, and again in 1818, it is stated that "previous to the discovery of vaccination the average number of deaths by small-pox within the (London) Bills of Mortality was 2,000 annually; whereas in the last year only 751 persons have died of the disease, although the increase of population within the last ten years has been 133,139."

The number 2,000 is about the average small-pox deaths of the whole eighteenth century, but those of the last two decades before the publication of Jenner's *Inquiry*, were 1,751 and 1,786, showing a decided fall. This, however, may pass. But when we come to the Report for 1826 we find the following: "But when we reflect that before the introduction of vaccination the average number of deaths from small-pox within the Bills of Mortality was annually about 4,000, no stronger argument can reasonably be demanded in favour of the value of this important discovery."

This monstrous figure was repeated in 1834, apparently quite forgetting the correct figure for the whole century given in 1818, and also the fact that the small-pox deaths recorded in the London Bills of Mortality in any year of the century never reached 4,000. But worse is to come; for in 1836 we have the following statement: "The annual loss of life by small-pox in the Metropolis, and within the Bills of Mortality only, before vaccination was established, exceeded 5,000, whereas in the course of last year only 300 died of the distemper." And in the Report for 1838 this gross error is repeated; while in the next year (1839) the conclusion is drawn "that 4,000 lives are saved every year in London since vaccination so largely superseded variolation." [1]

The Board of the National Vaccine Establishment consisted of the President and four Censors of the Royal College of Physicians, and the Master and two senior Wardens of the College of Surgeons. We cannot possibly suppose that they knew or believed that they were publishing untruths and grossly deceiving the public. We must, therefore, fall back upon the supposition that they were careless to such an extent

[1] These extracts from the Reports are given by Mr. White in his *Story of a Great Delusion*. The actual deaths from small-pox during the last century are given in the Second Report of the Royal Commission, p. 290. The extracts have been verified at the British Museum by my friend Dr. Scott Tebb, and are verbally accurate.

as not to find out that they were authorizing successive statements of the same quantity, as inconsistent with each other as 2,000 and 5,000.

The next example is given by Dr. Lettsom, who, in his evidence before the Parliamentary Committee in 1802, calculated the small-pox deaths of Great Britain and Ireland before vaccination at 36,000 annually; by taking 3,000 as the annual mortality in London and multiplying by twelve, because the population was estimated to be twelve times as large. He first takes a number which is much too high, and then assumes that the mortality in the town, village, and country populations was the same as in overcrowded, filthy London! Small-pox was always present in London, while Sir Gilbert Blane tells us that in many parts of the country it was quite unknown for periods of twenty, thirty, or forty years. In 1782 Mr. Connah, a surgeon at Seaford, in Sussex, only knew of one small-pox death in eleven years among a population of 700. Cross, the historian of the Norwich epidemic in 1819, states that previous to 1805 small-pox was little known in this city of 40,000 inhabitants, and was for a time almost extinct; and yet this gross error, of computing the small-pox mortality of the whole country from that of London (and computing it from wrong data) was not only accepted at the time, but has been repeated again and again down to the present day as an ascertained fact!

In a speech in Parliament in defence of vaccination, Sir Lyon Playfair gave 4,000 per million as the average London death-rate by small-pox before vaccination—a number nearly double that of the last twenty years of the century, which alone affords a fair comparison. But far more amazing is the statement by the late Dr. W. B. Carpenter, in a letter to the *Spectator* of April, 1881, that "a hundred years ago the small-pox mortality of London alone, with its then population of under a million, was often greater in a six months' epidemic than that of the twenty millions of England and Wales now is in any whole year." The facts, well known to every enquirer, are,—that the very highest small-pox mortality in the last century in a year was 3,992 in 1772, while in 1871 it was **7,912** in London, or more than double; and in the same year, in England and Wales, it was **23,000**. This amazing and almost incredible misstatement was pointed out and acknowledged privately, but never withdrawn publicly!

The late Mr. Ernest Hart, a medical man, editor of the *British Medical Journal*, and a great authority on sanitation, in his work entitled *The Truth about Vaccination*, surpassed even Dr. Carpenter in the monstrosity of his errors. At page 35 of the first edition (1880), he stated that in the forty years 1728–57 and 1771–80, the average annual small-pox mortality of London was about 18,000 per million living. The actual average mortality, from the tables given in the Second Report of the Royal Commission, page 290, was a little over 2,000, the worst periods having been chosen; and taking the lowest estimates of the population at the time, the mortality per million would have been under 3,000. This great authority, therefore, has multiplied the real number by six! In a later edition this statement is omitted; but in the first edition it was no mere misprint, for it was triumphantly dwelt upon over a whole page and compared with modern rates of mortality.

Yet one more official misstatement. About the year

1884 the National Health Society, with the approval of the Local Government Board, issued a tract entitled *Facts concerning Vaccination for Heads of Families*, in which appeared the statement, "*Before the introduction of vaccination, small-pox killed* 40,000 *persons yearly in this country*." We have already shown that Dr. Lettsom's figure, 36,000, was utterly unfounded, and probably three or four times greater than the truth. Here we have a semi-official and widely-distributed statement even more remote from the truth. In later issues of the same tract this particular statement is withdrawn, and a different but equally erroneous one substituted. Thus: "Before its discovery (vaccination) *the mortality from small-pox in London was forty times greater than it is now*." This is an altogether vague and misleading statement. If it means that in *some* years of the last century it was forty times greater than in *some* years of this century, it is misleading, because even within the last thirty years some years have a mortality not only forty but eighty and even 200 times as great as others. (In 1875 there were ten deaths per million, while in 1871 there were 2,420 deaths per million.) If it means on an average of say twenty years, it is false. For the twenty years 1869–98 the mortality was about 300 per million, while for the last twenty years before the discovery of vaccination it was about 2,000 per million, or less than *seven* times as much instead of *forty* times!

This same tract is full of other equally gross misstatements. It tells us, in large, black type, "*With due care in the performance of the operation, no risk of any injurious effects from it need be feared*." The

Registrar-General himself shows us that this is false, in his Report for 1895, Table 17, p. lii.:

COW-POX AND OTHER EFFECTS OF VACCINATION

| Year. | Deaths. | Year. | Deaths. |
|---|---|---|---|
| 1881 | 58 | 1889 | 58 |
| 1882 | 65 | 1890 | 43 |
| 1883 | 55 | 1891 | 43 |
| 1884 | 53 | 1892 | 58 |
| 1885 | 52 | 1893 | 59 |
| 1886 | 45 | 1894 | 50 |
| 1887 | 45 | 1895 | 56 |
| 1888 | 45 | | |

An average of 52 children officially murdered every year, and officially acknowledged, is termed "alleged injury," which need not be feared! And these cruel falsehoods are spread broadcast over the country, and the tract bears upon its title-page—"*Revised by the Local Government Board, and issued with their sanction.*"

As the tract bears no date, I cannot tell whether it is still issued; but it was in circulation up to the time when the Commission was sitting, and it is simply disgraceful that a Government Department should ever have given its official sanction to such a tissue of misrepresentations and palpable false statements. For these 785 deaths in fifteen years, and 390 in the preceding twenty-two years (classed as from erysipelas after vaccination), no one has been punished, and no compensation or even official apology has been given to the thousand sorrowing families. And we may be sure that these acknowledged deaths are only a small portion of what have really occurred, since the numbers have increased considerably in the later period, during which more attention has been given to such deaths and more inquests held. It is certain that for every such death acknowledged by the medical man concerned, many are concealed under the easy method of stating some of the later symptoms as the cause of death. Thus, Mr. Henry May, Medical Officer of Health, candidly states as follows: "In certificates given by us voluntarily, and to which the public have access, it is scarcely to be expected that a medical man will give opinions which may tell against or reflect upon himself in any way. In such cases he will most likely tell the truth, but not the whole truth, and assign some prominent symptom of the disease as the cause of death. As instances of cases which may tell against the medical man himself, I will mention erysipelas from vaccination, and puerperal fever. A death from the first cause occurred not long ago in my practice; and although I had not vaccinated the child, yet, in my desire to preserve vaccination from reproach, I omitted all mention of it from my certificate of death." (See *Birmingham Medical Review*, Vol III. pp. 34, 35.) That such *suppressio veri* is no new thing, but has been going on during the whole period of vaccination, is rendered probable by a statement in the *Medical Observer* of 1810, by Dr. Maclean. He says: "Very few deaths from cow-pox appear in the Bills of Mortality, owing to the means which have been used to suppress a knowledge of them. Neither were deaths, diseases, and failures transmitted in great abundance from the country, not because they did not happen, but because some practitioners were interested in not seeing them, and others who did see them were afraid of announcing what they knew."

As an example of the number of cases occurring all over the country, Mr. Charles Fox, a medical man residing at Cardiff, has published fifty-six cases of illness following vaccination, of which seventeen resulted in death (E. W. Allen, 1890). In only two of these, where he himself gave the certificate, was vaccination mentioned. All of these cases were examined by himself personally. Among those who survived, several were permanently injured in health, and some were crippled for life; while in most of such cases the inflammation and eruptions are so painful, and the sufferings of the children so great and so prolonged, that the mother endures continuous mental torture, lasting for weeks, months, or even years. And if *one* medical man can record such a mass of injury and disease in which vaccination was the palpable starting-point and certainly a contributory cause, what must be the total mass of unrecorded suffering throughout the whole country? Considering this and other evidence, together with the admitted and very natural concealment by the doctors concerned, "to save vaccination from reproach," the estimate of Mr. Alfred Milnes, a statistician who has paid special attention to the subject, that the officially admitted deaths must be at least multiplied by twelve to obtain the real deaths from vaccination, we shall arrive at the terrible number of over 600 children and adults killed annually by this compulsory operation; while, judging from the proportion of permanent injury, twenty-eight in Mr. Fox's fifty-six cases, with seventeen deaths, about 1,000 persons annually must suffer from it throughout their lives! As confirmatory of even this large amount, the testimony of Mr. Davidson, Medical Officer of Health for Congleton, and formerly a Public Vaccinator, is important. He began an inquiry into the alleged injurious effects of vaccination, without believing that they were serious. The outcome of his investigation was startling to him. In his Annual Report for 1893, he says: "In the investigation of a single vaccination period, the fact was revealed that in quite fifty per cent. of all vaccinated in that period (about seventy), the results were abnormal, and, in a large number of these very grave injuries had been inflicted. That the results of the practice are the same elsewhere as in Congleton I have no reason to doubt, for judging from what I have seen of his method of vaccinating, our Public Vaccinator is as careful as it seems possible for a Public Vaccinator to be."

This evidence of Mr. Davidson is especially important, because it reveals the fact that, as I stated some pages back, neither Public Vaccinators nor ordinary medical men usually know anything of the injurious effects of vaccination, except in such individual cases as may occur in their practice, while all around them there may be a mass of evil results which, when systematically investigated, proves as unexpected as it is startling in its amount.

This brief exposition of medical and official misstatements of facts and figures, always in favour of vaccination, might have been largely increased; but it is already sufficient to demonstrate the position I take, which is, that in this matter of Official and Compulsory Vaccination, both doctors and Government officials, however highly placed, however eminent, however honourable, are yet utterly untrustworthy. Beginning in the early years of the century, and continuing to our own times, we find the most gross and palpable blunders in figures—but always on the side of vaccination—and, on the testimony of medical men themselves, a more or less continuous perversion of the official records of vaccinal injury "in order to save vaccination from reproach." Let this always be remembered in any discussion of the question. The facts and figures of the medical profession, and of Government officials, in regard to the question of vaccination, *must never be accepted without verification.* And when we consider that these misstatements, and concealments, and denials of injury, have been going on throughout the whole of the century; that penal legislation has been founded on them; that homes of the poor have been broken up; that thousands have been harried by police and magistrates, have been imprisoned and treated in every way as felons; and that, at the rate now officially admitted, a thousand children have been certainly killed by vaccination during the last twenty years, and an unknown but probably much larger number injured for life, we are driven to the conclusion that those responsible for these reckless misstatements and their terrible results have, thoughtlessly and ignorantly, but none the less certainly, been guilty of a crime against liberty, against health, and against humanity, which will, before many years have passed, be universally held to be one of the foulest blots on the civilization of the nineteenth century.[1]

[1] As an example of the dreadful results of vaccination, even where special care was taken, the following case from the Sixth Report of the Royal Commission (p. 128) is worthy of earnest attention. It is the evidence of Dr. Thomas Skinner, of Liverpool:

*Q.* 20,766. Will you give the Commission the particulars of the case?—A young lady, fifteen years of age, living at Grove Park, Liverpool, was revaccinated by me at her father's request, during an outbreak of small-pox in Liverpool in 1865, as I had revaccinated all the girls in the Orphan Girls' Asylum in Myrtle Street, Liverpool (over 200 girls, I believe), and as the young lady's father was chaplain to the asylum, he selected, and I approved of the selection, of a young girl, the picture of health, and whose vaccine vesicle was matured, and as perfect in appearance as it is possible to conceive. On the eighth day I took off

## II

### MUCH OF THE EVIDENCE ADDUCED FOR VACCINATION IS WORTHLESS

WE will now proceed to discuss the alleged value of vaccination, by means of the best and widest statistical

the lymph in a capillary glass tube, almost filling the tube with clear, transparent lymph. Next day, 7th March, 1865, I re-vaccinated the young lady from this same tube, and from the same tube and at the same time I revaccinated her mother and the cook. Before opening the tube I remember holding it up to the light and requesting the mother to observe how perfectly clear and homogeneous, like water, the lymph was, neither pus nor blood corpuscles were visible to the naked eye. All three operations were successful, and on the eighth day all three vesicles were matured "like a pearl upon a rose petal," as Jenner described a perfect specimen. On that day, the eighth day after the operation, I visited my patient, and to all appearance she was in the soundest health and spirits, with her usual bright eyes and ruddy cheeks. Although I was much tempted to take the lymph from so healthy a vesicle and subject, I did not do so, as I have frequently seen erysipelas and other bad consequences follow the opening of a matured vesicle. As I did not open the vesicle that operation could not be the cause of what followed. Between the tenth and the eleventh day after the revaccination —that is, about three days after the vesicle had matured and begun to scab over—I was called in haste to my patient the young lady, whom I found in one of the most severe rigors I ever witnessed, such as generally precedes or ushers in surgical, puerperal, and other forms of fever. This would be on the 18th March, 1865. Eight days from the time of this rigor my patient was dead, and she died of the most frightful form of blood poisoning that I ever witnessed, and I have been forty-five years in the active practice of my profession. After the rigor, a low form of acute peritonitis set in, with incessant vomiting and pain, which defied all means to allay. At last stercoraceous vomiting, and cold, clammy, deadly sweats of a sickly odour set in, with pulselessness, collapse, and death, which closed the terrible scene on the morning of the 26th March, 1865. Within twenty minutes of death rapid decomposition set in, and within

evidence at our command; and in doing so we shall be able to show that the medical experts, who have been trusted by the Government and by the general public, are no less deficient in their power of drawing accurate conclusions from the official statistics of vaccination and small-pox mortality than they have been shown to be in their capacity for recording facts and quoting figures with precision and correctness.

In the elaborate paper by Sir John Simon, on the History and Practice of Vaccination, presented to Parliament in 1857 and reprinted in the First Report of the Royal Commission, he tells us that the earlier evidence of the value of vaccination was founded on *individual cases*, but that now "from individual cases the appeal is to masses of national experience." And

two hours so great was the bloated and discoloured condition of the whole body, more especially of the head and face, that there was not a feature of this once lovely girl recognisable. Dr. John Cameron, of 4, Rodney Street, Liverpool, physician to the Royal Southern Hospital at Liverpool, met me daily in consultation while life lasted. I have a copy of the certificate of death here.

*Q.* 20,767. To what do you attribute the death there?—I can attribute the death there to nothing but vaccination.

In the same Report, fifteen medical men give evidence as to disease, permanent injury, or death caused by vaccination. Two give evidence of syphilis and one of leprosy as clearly due to vaccination. And, as an instance of how the law is applied in the case of the poor, we have the story told by Mrs. Amelia Whiting (*QQ.* 21,434-21,464). To put it in brief, it amounts to this:—Mrs. Whiting lost a child, after terrible suffering, from inflammation supervening upon vaccination. The doctor's bill for the illness was £1 12*s.* 6*d.*; and a woman who came in to help was paid 6*s.* After this first child's death, proceedings were taken for the non-vaccination of another child; and though the case was explained in court, a fine of one shilling was inflicted. And through it all, the husband's earnings as a labourer were 11*s.* a week.

the marginal reference is, "Evidence on the protectiveness of vaccination must now be statistical." If this was true in 1857, how much more must it be so now, when we have forty years more of "national experience" to go upon. Dr. Guy, M.D., F.R.S., enforces this view in his paper published by the Royal Statistical Society in 1882. He says: "Is vaccination a preventive of small-pox? To this question there is, there can be, no answer except such as is couched in the language of figures." But the language of figures, otherwise the science of statistics, is not one which he who runs may read. It is full of pitfalls for the unwary, and requires either special aptitude or special training to avoid these pitfalls and deduce from the mass of figures at our command what they really teach.

A commission or committee of enquiry into this momentous question should have consisted wholly, or almost wholly, of statisticians, who would hear medical as well as official and independent evidence, would have all existing official statistics at their command, and would be able to tell us, with some show of authority, exactly what the figures proved, and what they only rendered probable on one side and on the other. But instead of such a body of experts, the Royal Commission, which for more than six years was occupied in hearing evidence and cross-examining witnesses, consisted wholly of medical men, lawyers, politicians, and country gentlemen, none of whom were trained statisticians, while the majority came to the enquiry more or less prejudiced in favour of vaccination. The report of such a body can have but little value, and I hope to satisfy my readers that it (the Majority Report) is not in accordance with the facts;

that the reporters have lost themselves in the mazes of unimportant details; and that they have fallen into some of the pitfalls which encumber the path of those who, without adequate knowledge or training, attempt to deal with great masses of figures.

But before proceeding to discuss the statistical evidence set forth in the reports of the Commission, I have again the disagreeable task of showing that a very large portion of it, on which the Commissioners mainly rely to justify their conclusions, is altogether untrustworthy, and must therefore be rejected whenever it is opposed to the results of the great body of more accurate statistical evidence. I allude of course to the question of the comparative small-pox mortality of the VACCINATED and the UNVACCINATED. The first point to be noticed is, that existing official evidence of the greatest value has never been made use of for the purposes of registration, and is not now available. For the last sixteen years the Registrar-General gives the deaths from small-pox under three headings. Thus, in the year 1881 he gives for London (Annual Summary, p. xxiv.):

| | | |
|---|---|---|
| Small-pox. | Vaccinated . . . | 524 deaths. |
| ,, | Not vaccinated . . | 962 ,, |
| ,, | No statement . . . | 885 ,, |

And in the year 1893, for England and Wales, the figures are (Annual Report, p. xi.):

| | | |
|---|---|---|
| Small-pox. | Vaccinated . . . | 150 deaths. |
| ,, | Unvaccinated . . . | 253 ,, |
| ,, | No statement . . . | 1054 ,, |

Now such figures as these, even if those under the first two headings were correct, are a perfect farce, and are totally useless for any statistical purpose. Yet every vaccination is officially recorded—since 1873 private as well as public vaccinations—and it would not have been difficult to trace almost every small-pox patient to his place of birth, and to get the official record of his vaccination if it exists. As the medical advisers of the Government have not done this, and give us instead partial and local statistics, usually under no official sanction and often demonstrably incorrect, every rule of evidence and every dictate of common sense entitle us to reject the fragmentary and unverified statements which they put before us. Of the frequent untrustworthiness of such statements it is necessary to give a few examples.

In *Notes on the Small-pox Epidemic at Birkenhead*, 1877 (p. 9), Dr. F. Vacher says: "Those entered as not vaccinated were admittedly unvaccinated, or without the faintest mark. The mere assertions of patients or their friends that they were vaccinated counted for nothing." Another medical official justifies this method of making statistics, as follows:—"I have always classed those as 'unvaccinated,' when no scar, presumably arising from vaccination, could be discovered. Individuals are constantly seen who state that they have been vaccinated, but upon whom no cicatrices can be traced. In a prognostic and a statistic point of view, it is better, and, I think, necessary, to class them as unvaccinated" (Dr. Gayton's Report for the Homerton Hospital for 1871–2–3).

The result of this method, which is certainly very general though not universal, is such a falsification of the real facts as to render them worthless for statistical purposes. It is stated by so high an authority as Sir James Paget, in his lectures on Surgical Pathology, that "cicatrices may in time wear out"; while the Vaccination Committee of the Epidemiological Society, in its Report for 1885–6, admitted that "not every cicatrice will permanently exist." Even more important is the fact, that in confluent small-pox the cicatrices are hidden, and large numbers of admissions to the hospitals are in the later stages of the disease. Dr. Russell, in his Glasgow Report (1871–2, p. 25), observes, "Sometimes persons were said to be vaccinated, but no marks could be seen, very frequently because of the abundance of the eruption. In some of those cases which recovered, an inspection before dismission discovered vaccine marks sometimes very good."

In many cases private enquiry has detected errors of this kind. In the Second Report of the Commission, pp. 219–20, a witness declared that out of six persons who died of small-pox and were reported by the medical officer of the Union to have been unvaccinated, five were found to have been vaccinated, one being a child who had been vaccinated by the very person who made the report, and another a man who had been twice revaccinated in the militia (*Q.* 6730–42). One other case may be given. In October, 1883, three unvaccinated children were stated in the Registrar-General's weekly return of deaths in London to have died of small-pox, "being one, four, and nine years of age, and all from 3, Medland Street, Stepney." On enquiry at the address given (apparently by oversight in this one case) the mother stated that the three children were hers, and that "all had been beautifully vaccinated." This case was investigated by Mr. J. Graham Spencer, of 33, Rigault Road, Fulham Park Gardens, and the facts were published in the local papers and also in *The Vaccination Inquirer* of December, 1883.

Several other cases were detected at Sheffield, and were adduced by Mr. A. Wheeler in his evidence before the Commission (6th Report, p. 70); and many others are to be found throughout the Anti-Vaccination periodicals. But the difficulty of tracing such misstatements is very great, as the authorities almost always refuse to give information as to the cases referred to when particular deaths from small-pox are recorded as "unvaccinated." Why this effort at secrecy in such a matter if there is nothing to hide? Surely it is to the public interest that official statistics should be made as correct as possible; and private persons who go to much trouble and expense in order to correct errors should be welcomed as public benefactors and assisted in every way, not treated as impertinent intruders on official privacy, as is too frequently the case.

The result of this prejudiced and unscientific method of registering small-pox mortality, is the belief of the majority of the medical writers on the subject that there is an enormous difference between the mortality of the vaccinated and the unvaccinated, and that the difference is due to the fact of vaccination or the absence of it. The following are a few of the figures as to this point given in the Reports of the Royal Commission:

| Authority. | Death Rate of Vaccinated. | Death Rate of Unvaccinated. |
|---|---|---|
| Dr. Gayton, in 2nd Report (Table B, p. 245). | 7·45 | 43 |
| Dr. Barry (Table F, p. 249) . . . . . | 8·1 | 32·7 |
| Sir John Simon (1st Rep., p. 74) . . | 0 to 12½ | 14½ to 60 |
| Mr. Sweeting, M.R.C.S. (2nd Rep., p. 119) . | 8·92 | 46·08 |

Now an immense body of statistics of the last century compiled by disinterested persons who had no interest to serve by making the severity of small-pox large or small, gives an average of from 14 to 18 per cent.[1] as the proportion of small-pox deaths to cases; and we naturally ask, How is it that, with so much better sanitary conditions and greatly improved treatment, nearly half the unvaccinated patients die, while in the last century less than one-fifth died? Many of the supporters of vaccination, such as Dr. Gayton (2nd Rep., p. 1856), have no explanation to offer. Others, such as Dr. Whitelegge (6th Rep., p. 533), believe that small-pox becomes more virulent periodically, and that one of its maxima of virulence caused the great epidemic of 1870–72, which, after more than half a century of vaccination, equalled some of the worst epidemics of the pre-vaccination period.

It is, however, a most suggestive fact that, considering small-pox mortality *per se*, without reference to vaccination—the records of which are, as have been shown, utterly untrustworthy—we find the case-mortality to agree closely with that of the last century. Thus, the figures given in the Reports of the Hampstead, Homerton, and Deptford small-pox hospitals at periods between 1876 and 1879, were, 19, 18·8, and 17 per cent. respectively (3rd Report, p. 205). If we admit that only the worst cases went to the hospitals, but also allow something for better treatment now, the result is quite explicable; whereas the other result, of a greatly *increased* fatality in the unvaccinated so exactly balanced by an alleged greatly *diminished* fatality in the vaccin-

[1] See Table J, p. 201, 3rd Report, and the Minority Report of the Roy. Comm., pp. 176–7.

ated is not explicable, especially when we remember that this diminished fatality applies to all ages, and it is now almost universally admitted that the alleged protective influence of vaccination dies out in ten or twelve years. These various opinions are really self-destructive. If epidemic small-pox is now much more virulent than in the last century, as shown by the greater mortality of the unvaccinated now than then, the greatly diminished or almost vanishing effect of primary vaccination in adults cannot possibly have reduced *their* fatality to one-fifth or one-sixth of that of the other class.

Again, it is admitted by many pro-vaccinist authorities that the unvaccinated, as a rule, belong to the poorer classes, while they also include most of the criminal classes, tramps, and generally the nomad population. They also include all those children whose vaccination has been deferred on account of weakness, or of their suffering from other diseases, as well as all those under vaccination age. The unvaccinated as a class are therefore especially liable to zymotic disease of any kind, small-pox included; and when, in addition to these causes of a higher death-rate from small-pox, we take account of the proved untrustworthiness of the statistics, wholly furnished by men who are prejudiced in favour of vaccination (as instanced by the declaration of Dr. Gayton, that when the eruption is so severe as on the third day to hide the vaccination marks, it affords *primâ facie* evidence of non-vaccination (2nd Report, Q. 1790), we are fully justified in rejecting all arguments in favour of vaccination supported by such fallacious evidence. And this is the more rational course to be adopted by all unprejudiced enquirers, because, as I shall now proceed to show, there is an abundance of facts of a more accurate and more satisfactory nature by which to test the question.[1]

One more point may be referred to before quitting this part of the subject, which is, that the more recent official hospital-statistics themselves afford a demonstration of the non-protective influence of vaccination, and thus serve as a complete refutation of the conclusions drawn from the statistics we have just been dealing with. Dr. Munk stated before the Hospital Commission, that the percentage of vaccinated patients in the London small-pox hospital had increased from 40 per cent. in 1838 to $94\frac{6}{10}$ per cent. in 1879 (3rd Report of Royal Comm., Q. 9090). This evidence was given in 1882; but Mr. Wheeler stated that, according to the Reports of the Highgate hospital, the vaccinated patients had long been over 90 per cent. of the whole, and are now often even 94 or 95 per cent. The hospitals of the Metropolitan Asylums Board, which take in mostly pauper patients, give a lower percentage—the Homerton hospital 85 per cent., the Deptford hospital 87 per cent., and the Hampstead hospital 75 per cent.—in the two latter cases adding the "doubtful" class to the vaccinated, as the facts already given prove that we have a right to do and still probably give too high a proportion of unvaccinated. As the proportion of the London population that is vaccinated cannot be over 90 per cent. (see Minority Report, pp. 173–4), and is probably

[1] The same view is taken even by some advocates of vaccination in Germany. In an account of the German *Commission for the Consideration of the Vaccination Question* in the *British Medical Journal*, August 29, 1885 (p. 408), we find it stated: "In the view of Dr. Koch, no other statistical material than the mortality from small-pox can be relied upon; questions as to the vaccinated or unvaccinated condition of the patient leaving too much room for error."

much lower, and considering the kind of patients the unvaccinated include (see back, p. 241), there remains absolutely nothing for the effects of vaccination. We have already seen that the total case-mortality of these hospitals agrees closely with that of the last century; the two classes of facts taken together thus render it almost certain that vaccination has never saved a single human life.

## III

### THE GENERAL STATISTICS OF SMALL-POX MORTALITY IN RELATION TO VACCINATION

HAVING thus cleared away the mass of doubtful or erroneous statistics depending on comparisons of the vaccinated and the unvaccinated in limited areas or selected groups of patients, we turn to the only really important evidence, those "masses of national experience" which Sir John Simon, the great official advocate of vaccination, tells us we must now appeal to for an authoritative decision on the question of the value of vaccination; to which may be added certain classes of official evidence serving as test cases or "control experiments" on a large scale. Almost the whole of the evidence will be derived from the Reports of the recent Royal Commission.

In determining what statistics really mean the graphic is the only scientific method, since, except in a few very simple cases, long tables of figures are confusing; and if divided up and averages taken, as is often done, they can be manipulated so as to conceal their real teaching. Diagrams, on the other hand, enable us to see the whole bearing of the variations that occur, while for comparison of one set of figures with another their superiority is overwhelming. This is especially the case with the statistics of epidemics and of general mortality, because the variations are so irregular and often so large as to render tables of figures very puzzling, while any just comparison of several tables with each other becomes impossible. I shall therefore put all the statistics I have to lay before my readers in the form of diagrams, which, I believe, with a little explanation, will enable any one to grasp the main points of the argument. (See end of volume.)

### *London Mortality and Small-pox*

The first and largest of the diagrams illustrating this question is that exhibiting the mortality of London from the year 1760 down to the present day (see end of volume). It is divided into two portions, that from 1760 to 1834 being derived from the old "Bills of Mortality," that from 1838 to 1896 from the Reports of the Registrar-General.

The "Bills of Mortality" are the only material available for the first period, and they are far inferior in accuracy to the modern registration, but they are probably of a fairly uniform character throughout, and may therefore be as useful for purposes of comparison as if they were more minutely accurate. It is admitted that they did not include the whole of the deaths, and the death-rates calculated from the estimated population will therefore be too low as compared with those of the Registrar-General, but the *course* of each death rate—its various risings or fallings—will probably be nearly true.[1] The years are given along

[1] It is always stated that only the deaths of those persons belonging to the Church of England, or who were buried in the churchyards, are recorded in the "Bills." This seems very improbable, because the "searchers" must have visited the house

the bottom of the diagram, and the deaths per million living are indicated at the two ends and in the centre, the last four years of the Bills of Mortality being omitted because they are considered to be especially inaccurate. The upper line gives the total death-rate from all causes, the middle line the death-rate from the chief zymotic diseases—measles, scarlet-fever, diphtheria, whooping-cough and fevers generally, excluding small-pox, and the lower line small-pox only. The same diseases, as nearly as they can be identified in the Bills of Mortality, according to Dr. Creighton, are given in the earlier portion of the diagram from the figures given in his great work, *A History of Epidemics in Britain*. As regards the line of small-pox mortality, the diagram is the same as that presented to the Royal Commission (3rd Report, diagram J), but it is carried back to an earlier date.

Let us now examine the lowest line, showing the small-pox death-rate. First taking the period from 1760 to 1800, we see, amid great fluctuations and some exceptional epidemics, a well-marked steady decline which, though obscured by its great irregularity, amounts to a difference of 1,000 per million living. This decline continues, perhaps somewhat more rapidly,

and recorded the death before the burial; and as they were of course paid a fee for each death certified by them, they would not enquire very closely as to the religious opinions of the family, or where the deceased was to be buried. A friend of mine who lived in London before the epoch of registration informs me that he remembers the "searchers'" visit on the occasion of the death of his grandmother. They were two women dressed in black; the family were strict dissenters, and the burial was at the Bunhill Fields cemetery for Nonconformists. This case proves that in all probability the "Bills" did include the deaths of many, perhaps most, Nonconformists.

to 1820. From that date to 1834 the decline is much less, and is hardly perceptible. The period of Registration opens with the great epidemic of 1838, and thenceforward to 1885 the decline is very slow indeed; while, if we average the great epidemic of 1871 with the preceding ten years, we shall not be able to discover any decline at all. From 1886, however, there is a rather sudden decline to a very low death-rate, which has continued to the present time. Now it is alleged by advocates of vaccination, and by the Commissioners in their Report, that the decline from 1800 onwards is due to vaccination, either wholly or in great part, and that "the marked decline of small-pox in the first quarter of the present century affords substantial evidence in favour of the protective influence of vaccination."[1] This conclusion is not only entirely unwarranted by the evidence on any accepted methods of scientific reasoning, but it is disproved by several important facts. In the first place the decline in the first quarter of the century is a clear continuation of a decline which had been going on during the preceding forty years, and whatever causes produced that earlier decline may very well have produced the continuation of it. Again, in the first quarter of the century, vaccination was comparatively small in amount and imperfectly performed. Since 1854 it has been compulsory and almost universal; yet from 1854 to 1884 there is almost no decline of small-pox perceptible, and the severest epidemic of the century occurred in the midst of that period. Yet again, the one clearly marked decline of small-pox has been in the ten years from 1886 to 1896, and it is precisely in this period that there has been a great falling off in vaccination in

[1] *Final Report of Roy. Comm.*, p. 20 (par. 85).

London, from only 7 per cent. less than the births in 1885 to 20·6 per cent. less in 1894, the last year given in the Reports of the Local Government Board; and the decrease of vaccinations has continued since.

But even more important, as showing that vaccination has had nothing whatever to do with the decrease of small-pox, is the very close general parallelism of the line showing the other zymotic diseases, the diminution of which it is admitted has been caused by improved hygienic conditions. The decline of this group of diseases in the first quarter of this century, though somewhat less regular, is quite as well marked as in the case of small-pox, as is also its decline in the last forty years of the 18th century, strongly suggesting that both declines are due to common causes. Let any one examine this diagram carefully and say if it is credible that from 1760 to 1800 both declines are due to some improved conditions of hygiene and sanitation, but that after 1800, while the zymotics have continued to decline from the same class of causes one zymotic—small-pox—*must* have been influenced by a new cause—vaccination, to produce its corresponding decline. Yet this is the astounding claim made by the Royal Commissioners! And if we turn to the other half of the diagram showing the period of registration, the difficulty becomes even greater. We first have a period from 1838 to 1870, in which the zymotics actually rose; and from 1838 to 1871, averaging the great epidemic with the preceding ten years, we find that small-pox also rose, or at the best remained quite stationary. From 1871 to 1875 zymotics are much lower, but run quite parallel with small-pox; then there is a slight decline in both, and zymotics and small-pox remain lower in the last ten years than they have ever been before, although in this last period vaccination has greatly diminished.

Turning to the upper line, showing the death-rate from all causes, we again find a parallelism throughout, indicating improved general conditions acting upon *all* diseases. The decline of the total death-rate from 1760 to 1810 is remarkably great, and it continues at a somewhat less rate to 1830, just as do the zymotics and small-pox. Then commences a period from 1840 to 1870 of hardly perceptible decline partly due to successive epidemics of cholera, again running parallel with the course of the zymotics and of small-pox; followed by a great decline to the present time, corresponding in amount to that at the beginning of the century.

The Commissioners repeatedly call attention to the fact that the mortality from measles has not at all declined and that other zymotics have not declined in the same proportion as small-pox, and they argue: "If improved sanitary conditions were the cause of small-pox becoming less, we should expect to see that they had exercised a similar influence over almost all other diseases. Why should they not produce the same effect in the case of measles, scarlet fever, whooping-cough, and indeed any disease spread by contagion or infection and from which recovery was possible?" This seems a most extraordinary position to be taken in view of the well-known disappearance of various diseases at *different* epochs. Why did leprosy almost disappear from England at so early a period and plague later on? Surely to *some* improved conditions of health. The Commissioners do not, and we may presume cannot, tell us why measles, of all the zymotic diseases, has rather increased than diminished during the whole of this century. Many students of epidemics hold that certain diseases are liable to replace each other, as suggested by Dr. Watt, of Glasgow, in the case of measles and small-pox. Dr. Farr, the great medical statistician, adopted this view. In his Annual Report to the Registrar-General in 1872 (p. 224), he says: "The zymotic diseases replace each other; and when one is rooted out it is apt to be replaced by others which ravage the human race indifferently whenever the conditions of healthy life are wanting. They have this property in common with weeds and other forms of life: as one species recedes another advances." This last remark is very suggestive in view of the modern germ-theory of these diseases. This substitution theory is adopted by Dr. Creighton, who in his *History of Epidemics in England* suggests that plague was replaced by typhus fever and small-pox; and, later on, measles, which was insignificant before the middle of the seventeenth century, began to replace the latter disease. In order to show the actual state of the mortality from these diseases during the epoch of registration, I have prepared a diagram (II.) giving the death-rates for London of five of the chief zymotics, from the returns of the Registrar-General, under the headings he adopted down to 1868—for to divide fevers into three kinds for half the period, and to separate scarlatina and diphtheria, as first done in 1859, would prevent any useful comparison from being made.

The lowest line, as in the larger diagram, shows Small-pox. Above it is Measles, which keeps on the whole a very level course, showing, however, the high middle period of the zymotics and two low periods, from 1869 to 1876, and from 1848 to 1856, the first nearly corresponding to the very high small-pox death-rate from 1870 to 1881; and the other just following the two small-pox epidemics of 1844 and 1848, thus supporting the view that it is in process of replacing that disease. Scarlatina and diphtheria show the high rate of zymotics generally from 1848 to 1870, with a large though irregular decline subsequently. Whooping-cough shows a nearly level course to 1882 and then a well-marked decline. Fevers (typhus, enteric, and simple) show the usual high middle period, but with an earlier and more continuous decline than any of the other zymotic diseases. We thus see that all these diseases exhibit common features though in very different degrees, all indicating the action of general causes, some of which it is by no means difficult to point out.

In 1845 began the great development of our railway system, and with it the rapid growth of London, from a population of two millions in 1844 to one of four millions in 1884. This rapid growth of population was at first accompanied with over-crowding, and as no adequate measures of sanitation were then provided the conditions were prepared for that increase of zymotic disease which constitutes so remarkable a feature of the London death-rates between 1848 and 1866. But at the latter date commenced a considerable decline both in the total mortality and in that from all the zymotic diseases, except measles and small-pox, but more especially in fevers and diphtheria, and this decrease is equally well explained by the completion, in 1865, of that gigantic work, the main drainage of London. The last marked decline in small-pox, in fevers, and to a less marked degree in whooping-cough, is coincident with a recognition of the fact that hospitals are themselves often centres of contagion, and the establishment of floating hospitals for London cases of small-pox. Perhaps even more beneficial was the modern system of excluding sewer-gas from houses.

We thus see that the increase or decrease of the chief zymotic diseases in London during the period of registration is clearly connected with adverse or favourable hygienic conditions of a definite kind. During the greater part of this period small-pox and measles alone showed no marked increase or decrease, indicating that the special measures affecting them had not been put in practice, till ten years back the adoption of an effective system of isolation in the case of small-pox has been followed by such marked results wherever it has been adopted as to show that *this* is the one method yet tried that has produced any large and unmistakable effect, thus confirming the experience of the town of Leicester, which will be referred to later on.

The Commissioners, in their *Final Report*, lay the greatest stress on the decline of small-pox at the beginning of the century, which "followed upon the introduction of vaccination," both in England, in Western Europe, and in the United States. They declare that "there is no proof that sanitary improvements were the main cause of the decline of small-pox," and that "no evidence is forthcoming to show that during the first quarter of the nineteenth century these improvements differentiated that quarter from the last quarter or half of the preceding century in any way at all comparable to the extent of the differentiation in respect to small-pox" (p. 19, par. 79). To the accuracy of these statements I demur in the strongest manner. There *is* proof that sanitary improvements were the main cause of this decline of small-pox early in the century, viz., that the other zymotic diseases as a whole showed a simultaneous decline to a nearly equal amount, while the general death-rate showed a decline to a much greater amount, both admittedly due to improved hygienic conditions, since there is *no other known cause* of the diminution of disease; and that the Commissioners altogether ignore these two facts affords, to my mind, a convincing proof of their incapacity to deal with this great statistical question. And, as to the second point, I maintain that there *is* ample direct evidence, for those who look for it, of great improvements in the hygienic conditions of London quite adequate to account for the great decline in the general mortality, and therefore equally adequate to account for the lesser declines in zymotic diseases and in small-pox, both of which began in the last century, and only became somewhat intensified in the first quarter of the present century, to be followed twenty years later by a complete check or even a partial rise. This rise was equally marked in small-pox as in the other diseases, and thus proved, as clearly as anything can be proved, that its decline and fluctuations are in no way dependent on vaccination, but are due to causes of the very same general nature as in the case of other diseases.

To give the evidence for this improvement in London hygiene would, however, break the continuity of the discussion as to small-pox and vaccination; but the comparison of the general and zymotic death-rates with that of small-pox, exhibits so clearly the identity of the causes which have acted upon them all as to render the detailed examination of the various improved conditions that led to the diminished mortality unnecessary. The diagram showing the death-rates from these three causes of itself furnishes a complete refutation of the Commissioners' argument. The evidence as to the nature of the improved conditions is given in an appendix at the end of this chapter.

### *Small-pox and other Diseases in Britain during the Period of Registration*

We have no general statistics of mortality in England and Wales till the establishment of the Registration system in 1838, but the results make up for their limited duration by their superior accuracy. Till the year 1870 no record was kept of the amount of vaccination, except as performed by the public vaccinators; but since 1872 all vaccinations are recorded, and the numbers published by the Local Government Board. My third diagram is for the purpose of showing graphically the relation of small-pox to other zymotic diseases, and to vaccination, for England and Wales. The lower line shows small-pox, the middle one zymotic diseases, and the upper the total death-rates. The relations of the three are much the same as in the London diagram, the beginning of the great decline of zymotics being in 1871, and that of small-pox in 1872, but the line of small-pox is much lower, and zymotics somewhat lower than in London, due to a larger proportion of the inhabitants living under comparatively healthy rural conditions.

But if the amount of vaccination were the main and almost exclusive factor in determining the amount of small-pox, there ought to be little or no difference between London and the country. But here, as in all other cases, the great factor of comparative density of population in compared areas, is seen to have its full effect on small-pox mortality as in that of all other zymotic diseases.

This non-relation between vaccination and small-pox mortality is further proved by the dotted line, showing the total vaccinations per cent. of births for the last 22 years, as given in the "Final Report" (p. 34). The diminution of vaccination in various parts of the country began about 1884, and from 1886 has been continuous and rapid, and it is during this very period that small-pox has been continuously less in amount than has ever been known before. Both in the relation of London small-pox to that of the whole country, and in the relation of small-pox to vaccination, we find proof of the total inefficacy of that operation.

### *Small-pox in Scotland and in Ireland*

In their *Final Report* the Commissioners give us Tables of the death-rates from small-pox, measles, and scarlet-fever in Scotland and Ireland; and from these Tables I have constructed my diagram (IV.), combining the two latter diseases for simplicity, and including the period of compulsory vaccination and accurate registration in both countries.

The most interesting feature of this diagram is the striking difference in the death-rates of the two countries. Scotland, the richer, more populous, and more prosperous country having a much greater mortality, both from the two zymotics and from small-pox, than poor, famine-stricken, depopulated Ireland. The maximum death-rate by the two zymotics in Scotland is considerably more than double that in Ireland, and the minimum is larger in the same proportion. In small-pox the difference is also very large in the same direction, for although the death-rate during the great epidemic in 1872 was only one-fourth greater in Scotland, yet as the epidemic there lasted three years, the total death-rate for those years was nearly twice as great as for the same period in Ireland, which, however, had a small epidemic later on in 1878. Since 1883 small-pox has been almost absent from both countries, as from England; but taking the twenty years of repeated epidemics from 1864 to 1883, we find the average small-pox death-rate of Scotland to be about 139, and that of Ireland 85 per million, or considerably more than as three to two. But even Scotland had a much lower small-pox mortality than England, the proportions being as follows for the three years which included the epidemic of 1871–3:

Ireland, 800 per million in the three years.
Scotland, 1,450 per million in the three years.
England, 2,000 per million in the three years.

Now the Royal Commissioners make no remark whatever on these very suggestive facts, and they have arranged the information in tables in such a way as to render it very difficult to discover them; and this is another proof of their incapacity to deal with statistical questions. They seem to be unable to look at small-pox from any other point of view than that of the vaccinationist, and thus miss the essential features of the evidence they have before them. Every statistician knows the enormous value of the representation of tabular statistics by means of diagrammatic curves. It is the only way by which in many cases the real teaching of statistics can be detected. An enormous number of such diagrams, more or less instructive and complete, were presented to them, and, at great cost, are printed in the Reports; but I cannot find that, in their *Final Report*, they have made any adequate use of them, or have once referred to them, and thus it is that they have overlooked so many of the most vital teachings of the huge mass of figures with which they had to deal.

It is one of the most certain of facts relating to sanitation, that comparative density of population affects disease, and especially the zymotic diseases, more than any other factor that can be ascertained. It is mainly a case of purity of the air, and consequent purification of the blood; and when we consider that breathing is the most vital and most continuous of all organic functions, that we must and do breathe every moment of our lives, that the air we breathe is taken into the lungs, one of the largest and most delicate organs of the body, and that the air so taken in acts directly upon the blood, and thus affects the whole organism, we see at once how vitally important it is that the air around us should be as free as possible from contamination, either by the breathing of other people, or by injurious gases or particles from decomposing organic matter, or by the germs of disease. Hence it happens that under our present terribly imperfect social arrangements the death-rate (other things being equal) is a function of the population per square mile, or perhaps more accurately of the proportion of town to rural populations.

In the light of this consideration let us again compare these diagrams of Irish, Scottish, and English death-rates. In Ireland only 11 per cent. of the population live in the towns of 100,000 inhabitants and upwards. In Scotland 30 per cent., and in England and Wales 54 per cent.; and we find the mortality from zymotic diseases to be roughly proportional to these figures. We see here unmistakable cause and effect. Impure air, with all else that overcrowding implies, on the one hand, higher death-rate on the other. This explains the constant difference between London and rural mortality, and it also explains what seems to have puzzled the Commissioners more than anything else—the intractability of some of the zymotics to ordinary sanitation, as in the case of measles especially, and in a less degree of whooping-cough—for in their case the continual growth of urban as opposed to rural populations has neutralised the effects of such improved conditions as we have been able to introduce.

But the most important fact for our present purpose is, that small-pox is subject to this law just as are the other zymotics, while it pays no attention whatever to vaccination. The statistician to the Registrar-General for Scotland gave evidence that ever since 1864 more than 96 per cent. of the children born have been vaccinated or had had previous small-pox, and he makes no suggestion of any deficiency that can be remedied. But in the case of Ireland the medical commissioner for the Local Government Board for Ireland, Dr. MacCabe, told the Commissioners that vaccination there was very imperfect, and that a large proportion of the population was "unprotected by vaccination," this state of things being due to various causes, which he explained (2nd Rep., QQ. 3,059–3,075). But neither Dr. MacCabe nor the Commissioners notice the suggestive, and from their point of view alarming, fact that imperfectly vaccinated Ireland had had far less small-pox mortality than thoroughly well-vaccinated Scotland, enormously less than well-vaccinated England, and overwhelmingly less than equally well-vaccinated London. Ireland—Scotland—England—London—a graduated series in density of population, and in zymotic death-rate; the small-pox death-rate increasing in the same order and to an enormous extent, quite regardless of the fact that the last three have had practically complete vaccination during the whole period of the comparison; while Ireland alone, with the lowest small-pox death-rate by far, has, on official testimony, the least amount of vaccination. And yet the majority of the Commissioners still pin their faith on vaccination, and maintain that the cumulative force of the testimony in its favour is irresistible! And further, that "sanitary improvements" cannot be asserted to afford "an adequate explanation of the diminished mortality from small-pox."

It will now be clear to my readers that these conclusions, set forth as the final outcome of their seven years' labours, are the very reverse of the true ones; and that they have arrived at them by neglecting altogether to consider, *in their mutual relations*, "those great masses of national statistics" which alone can be depended on to point out true causes, but have limited themselves to such facts as the alleged mortalities of the vaccinated and the unvaccinated, changes of age-incidence, and other matters of detail, some of which are entirely vitiated by untrustworthy evidence while others require skilled statistical treatment to arrive at true results, a subject quite beyond the powers of untrained physicians and lawyers, however eminent in their own special departments.[1]

[1] As an example of the Commissioners' statistical fallacies in

*Small-pox and Vaccination on the Continent*

Before proceeding to discuss those special test-cases in our own country which still more completely show the impotence of vaccination, it will be well to notice a few Continental States which have been, and still are, quoted as affording illustrations of its benefits.

We will first take Sweden, which has had fairly complete national statistics longer than any other country; and we are now fortunately able to give the facts on the most recent official testimony—the Report furnished by the Swedish Board of Health to the Royal Commission, and published in the Appendix to their Sixth Report (pp. 751–56). Such great authorities as Sir William Gull, Dr. Seaton, and Mr. Marson, stated before the Committee of Enquiry in 1871 that Sweden was one of the best vaccinated countries, and that the Swedes were the best vaccinators. Sir John Simon's celebrated paper, which was laid before Parliament in 1857 and was one of the chief supports of compulsory legislation, made much of Sweden, and had a special diagram to illustrate the effects of vaccination on small-pox. This paper is reproduced in the First Report of the recent Royal Commission (pp. 61–113), and we find the usual comparison of small-pox mortality in the last and present century which is held to be conclusive as to the benefits of vaccination. He says vaccination was introduced in 1801, and divides his diagram into two halves differently coloured before and after this date. It will be observed that, as in

treating the subject of changed age-incidence, see Mr. Alexander Paul's *A Royal Commission's Arithmetic* (King & Son, 1897), and, especially, Mr. A. Milnes' *Statistics of Small-pox and Vaccination* in the Journal of the Royal Statistical Society, September, 1897.

England, there was a great and sudden decrease of small-pox mortality after 1801, the date of the *first* vaccination in Sweden, and by 1812 the whole reduction of mortality was completed. But from that date for more than sixty years there was an almost continuous *increase* in frequency and severity of the epidemics. To account for this sudden and enormous decrease Sir John Simon states, in a note, and without giving his authority: "About 1810 the vaccinations were amounting to nearly a quarter of the number of births." But these were almost certainly both adults and children of various ages, and the official returns now given show that down to 1812, when the whole reduction of small-pox mortality had been effected, only 8 per cent. of the population had been vaccinated. We are told, in a note to the official tables, that the *first successful vaccination in Stockholm was at the end of 1810*, so that the earlier vaccinations must have been mainly in the rural districts; yet the earlier Stockholm epidemics in 1807, before a single inhabitant was vaccinated, and in 1825, were less severe than the six later ones, when vaccination was far more general.

Bearing these facts in mind, and looking at diagram V., we see that it absolutely negatives the idea of vaccination having had anything to do with the great reduction of small-pox mortality, which was almost all effected *before* the first successful vaccination in the capital on the 17th December, 1810! And this becomes still more clear when we see that, as vaccination increased among a population which, the official Report tells us, had the most "perfect confidence" in it, small-pox epidemics increased in virulence, especially in the capital (shown in the diagram by the dotted peaks) where, in 1874, there was a small-pox mortality of 7,916 per million, reaching 10,290 per million during the whole epidemic, which lasted two years. This was worse than the worst epidemic in London during the eighteenth century.[1]

But although there is no sign of a relation between vaccination and the decrease of small-pox, there is a very clear relation between it and the decrease in the general mortality. This is necessarily shown on a much smaller vertical scale to bring it into the diagram. If it were on the same scale as the small-pox line, its downward slope would be four times as rapid as it is. The decrease in the century is from about 27,000 to 15,000 per million, and, with the exception of the period of the Napoleonic wars, the improvement is nearly continuous throughout. There has evidently been a great and continuous improvement in healthy conditions of life in Sweden, as in our own country and probably in all other European nations; and this improvement, or some special portion of it, must have acted powerfully on small-pox to cause the enormous diminution of the disease down to 1812, with which, as we have seen, vaccination could have had nothing to do. The only thing that vaccination seems to have done is, to have acted as a check to this diminution, since it is otherwise impossible to explain the complete cessation of improvement as the operation became more general; and this is more especially the case in view of the fact that the general death-rate has continued to decrease at almost the same rate down to the present day!

[1] The highest small-pox mortality in London was in 1772, when 3,992 deaths were recorded in an estimated population of 727,000, or a death-rate of not quite 5,500 per million. (See Second Report, p. 290.)

The enormous small-pox mortality in Stockholm has been explained as the result of very deficient vaccination; but the Swedish Board of Health states that this deficiency was more apparent than real, first, because 25 per cent. of the children born in Stockholm die before completing their first year, and also because of neglect to report private vaccinations, so that "the low figures for Stockholm depend more on the cases of vaccination not having been reported than on their not having been effected." (Sixth Report, p. 754, 1st col., 3rd par.)

The plain and obvious teaching of the facts embodied in this diagram is, that small-pox mortality is in no way influenced (except it be injuriously) by vaccination, but that here, as elsewhere, it does bear an obvious relation to density of population; and also that, when uninfluenced by vaccination, it follows the same law of decrease with improved conditions of general health as does the total death-rate.

This case of Sweden alone affords complete proof of the uselessness of vaccination; yet the Commissioners in the *Final Report* (par. 59) refer to the great diminution of small-pox mortality in the first twenty years of the century as being due to it. They make no comparison with the total death-rate; they say nothing of the increase of small-pox from 1824 to 1874; they omit all reference to the terrible Stockholm epidemics increasing continuously for fifty years of legally enforced vaccination and culminating in that of 1874, which was far worse than the worst known in London during the whole of the eighteenth century. Official blindness to the most obvious facts and conclusions can hardly have a more striking illustration than the appeal to the case of Sweden as being favourable to the claims of vaccination.

My next diagram (No. VI.) shows the course of small-pox in Prussia since 1816, with an indication of the epidemics in Berlin in 1864 and 1871. Dr. Seaton, in 1871, said to the Committee on Vaccination (Q. 5,608), "I know Prussia is well protected," and the general medical opinion was expressed thus in an article in the *Pall Mall Gazette* (May 24, 1871): "Prussia is the country where revaccination is most generally practised, the law making the precaution obligatory on every person, and the authorities conscientiously watching over its performance. As a natural result, cases of small-pox are rare." Never was there a more glaring untruth than this last statement. It is true that revaccination was enforced in public schools and other institutions, and most rigidly in the Army, so that a very large proportion of the adult male population must have been revaccinated; but, instead of cases of small-pox being rare, there had been for the twenty-four years preceding 1871 a much *greater* small-pox mortality in Prussia than in England, the annual average being 248 per million for the former and only 210 for the latter. A comparison of the two diagrams shows the difference at a glance. English small-pox only once reached 400 per million (in 1852), while in Prussia it four times exceeded that amount. And immediately after the words above quoted were written, the great epidemic of 1871–72 caused a mortality in revaccinated Prussia more than double that of England! *Now*, after these facts have been persistently made known by the anti-vaccinators, the amount of vaccination in Prussia before 1871 is depreciated, and Dr. A. F. Hopkirk actually classes it among countries "without compulsory vaccination." (See table and diagram opposite p. 238 in the 2nd Report.)

In the city of Berlin we have indicated two epidemics, that in 1864, with a death-rate a little under 1,000 per million, while that in 1871 rose to 6,150 per million, or considerably more than twice as much as that of London in the same year, although the city must have contained a very large male population which had passed through the army, and had therefore been revaccinated.

I give one more diagram (No. VII.) of small-pox in Bavaria, from a table laid before the Royal Commission by Dr. Hopkirk for the purpose of showing the results of long-continued compulsory vaccination. He stated to the Commission that vaccination was made compulsory in 1807, and that in 1871 there were 30,742 cases of small-pox, of which 95·7 per cent. were vaccinated. (2nd Report, Q. 1,489.) He then explains that this was because "nearly the whole population was vaccinated"; but he does not give any figures to prove that the vaccinated formed more than this proportion of the whole population; and as the vaccination age was one year, it is certain that they did not do so.[1] He calls this being "slightly attacked," and argues that it implies "some special protection." No doubt the small-pox mortality of Bavaria was rather low, about equal to that of Ireland; but in 1871 it rose to over 1,000 per million, while Ireland had only 600, besides which the epidemic lasted for two years, and was therefore very nearly equal to that of England. But we have the explanation when we look at the line showing the other zymotics, for these are decidedly lower than those of England, showing better general

[1] The small-pox deaths under one year in England have varied during the last fifty years from 8·6 to 27 per cent. of the whole. (See *Final Report*, p. 151.)

sanitary conditions. In Bavaria, as in all the other countries we have examined, the behaviour of small-pox shows no relation to vaccination, but the very closest relation to the other zymotics and to density of population. The fact of 95·7 per cent. of the small-pox patients having been vaccinated agrees with that of our Highgate hospital, but is even more remarkable as applying to the population of a whole country, and is alone sufficient to condemn vaccination as useless. And as there were 5,070 deaths to these cases, the fatality was 16·5 per cent., or almost the same as that of the last century; so that here again, and on a gigantic scale, the theory that the disease is "mitigated" by vaccination, even where not prevented, is shown to be utterly baseless. Yet this case of Bavaria was chosen by a strong vaccinist as affording a striking proof of the value of vaccination when thoroughly carried out; and I cannot find that the Commissioners took the trouble to make the comparisons here given, which would at once have shown them that what the case of Bavaria really proves is the complete uselessness of vaccination.

This most misleading, unscientific, and unfair proceeding, of giving certain figures of small-pox mortality among the well vaccinated, and then, without any adequate comparison, *asserting* that they afford a proof of the value of vaccination, may be here illustrated by another example. In the original paper by Sir John Simon on the *History and Practice of Vaccination*, presented to Parliament in 1857, there is, in the Appendix, a statement by Dr. T. Graham Balfour, surgeon to the Royal Military Asylum for Orphans at Chelsea, as to the effects of vaccination in that institution—that since the opening of the Asylum in 1803 the Vaccination Register has been accurately kept, and that every one who entered was vaccinated unless he had been vaccinated before or had had small-pox; and he adds: "Satisfactory evidence can therefore, in this instance, be obtained that they were all protected." Then he gives the statistics, showing that during forty-eight years, from 1803 to 1851, among 31,705 boys there were thirty-nine cases and four deaths, giving a mortality at the rate of 126 per million on the average number in the Asylum, and concludes by saying: "The preceding facts appear to offer most conclusive proofs of the value of vaccination." But he gives no comparison with other boys of about the same age and living under equally healthy conditions, but who had not been so uniformly or so recently vaccinated; for it must be remembered that, as this was long before the epoch of compulsory vaccination, a large proportion of the boys would be unvaccinated at their entrance, and would therefore have the alleged benefit of a recent vaccination. But when we make the comparison, which both Dr. Balfour and Sir John Simon failed to make, we find that these well vaccinated and protected boys had a *greater* small-pox mortality than the imperfectly protected outsiders. For in the First Report of the Commission (p. 114, Table B) we find it stated that in the period of optional vaccination (1847-53) the death-rate from small-pox of persons from ten to fifteen years[1] was 94 per million! Instead of offering "most conclusive proofs of the value of vaccination," his own facts and figures, if they prove anything at all, prove

[1] This almost exactly agrees with the ages of the boys who are admitted between nine and eleven, and leave at fourteen. (See Low's *Handbook of London Charities.*)

not only the uselessness but the evil of vaccination, and that it really tends to *increase* small-pox mortality. And this conclusion is also reached by Professor Adolf Vogt, who, in the elaborate statistical paper sent by him to the Royal Commission, and printed in their Sixth Report, but not otherwise noticed by them, shows by abundant statistics from various countries, that the small-pox death-rate and fatality have been *increased* during epidemics occurring in the epoch of vaccination.

One more point deserves notice before leaving this part of the inquiry, which is, the specially high small-pox mortality of great commercial seaports. The following table, compiled from Dr. Pierce's *Vital Statistics* for the Continental towns and from the Reports of the Royal Commission for those of our own country, is very remarkable and instructive.

| Name of Town. | Year. | Small-pox Death-rate per Million. |
|---|---|---|
| Hamburgh . . . | 1871 | 15,440 |
| Rotterdam . . . | 1871 | 14,280 |
| Cork . . . . . | 1872 | 9,600 |
| Sunderland . . . | 1871 | 8,650 |
| Stockholm . . . | 1874 | 7,916 |
| Trieste . . . . | 1872 | 6,980 |
| Newcastle-on-Tyne . | 1871 | 5,410 |
| Portsmouth . . . | 1872 | 4,420 |
| Dublin . . . . | 1872 | 4,330 |
| Liverpool . . . | 1871 | 3,890 |
| Plymouth . . . | 1872 | 3,000 |

The small-pox death-rate in the case of the lowest of these towns is very much higher than in London during the same epidemic, and it is quite clear that vaccination can have had nothing to do with this difference. For if it be alleged that vaccination was neglected in Hamburgh and Rotterdam, of which we find no particulars, this cannot be said of Cork, Sunderland, and Newcastle. Again, if the very limited and imperfect vaccination of the first quarter of the century is to have the credit of the striking reduction of small-pox mortality that then occurred, as the Royal Commissioners claim, a small deficiency in the very much more extensive and better vaccination that generally prevailed in 1871, cannot be the explanation of a small-pox mortality *greater* than in the worst years of London when there was no vaccination. Partial vaccination cannot be claimed as producing marvellous effects at one time and less than nothing at all at another time, yet this is what the advocates of vaccination constantly do. But on the sanitation theory the explanation is simple. Mercantile seaports have grown up along the banks of harbours or tidal rivers whose waters and shores have been polluted by sewage for centuries. They are always densely crowded, owing to the value of situations as near as possible to the shipping. Hence there is always a large population living under the worst sanitary conditions, with bad drainage, bad ventilation, abundance of filth and decaying organic matter, and all the conditions favourable to the spread of zymotic diseases and their exceptional fatality. Such populations have maintained to our day the insanitary conditions of the last century, and thus present us with a similarly great small-pox mortality, without any regard to the amount of vaccination that may be practised. In this case they illustrate the same principle which so well explains the very different amounts of small-pox mortality in Ireland, Scotland, England, and London, with hardly any difference in the quantity of vaccination.

The Royal Commissioners, with all these facts before them or at their command, have made none of these comparisons. They give the figures of small-pox mortality, and either explain them by alleged increase or decrease of vaccination, or argue that, as some other disease—such as measles—did not decrease at the same time or to the same amount, therefore sanitation cannot have influenced small-pox. They never once compare small-pox mortality with general mortality, or with the rest of the group of zymotics, and thus fail to see their wonderfully close agreement—their simultaneous rise and fall, which so clearly shows their subjection to the same influences and proves that no special additional influence can have operated in the case of small-pox.

## IV

### TWO GREAT EXPERIMENTS WHICH ARE CONCLUSIVE AGAINST VACCINATION

Those who disbelieve in the efficacy of vaccination to protect against small-pox are under the disadvantage that, owing to the practice having been so rapidly adopted by all civilized people, there are no communities who have rejected it while adopting methods of general sanitation, and who have also kept satisfactory records of mortality from various causes. Any such country would have afforded what is termed a "control" or test experiment, the absence of which vitiates all the evidence of the so-called "variolous test" in Jenner's time, as was so carefully pointed out before the Commission by Dr. Creighton and Professor Crookshank. We do, however, now possess two such tests on a limited, but still a sufficient, scale. The first is that of the town of Leicester, which for the last twenty years has rejected vaccination till it has now almost vanished altogether. The second is that of our Army and Navy, in which, for a quarter of a century, every recruit has been revaccinated, unless he has recently been vaccinated or has had small-pox. In the first we have an almost wholly "unprotected" population of nearly 200,000, which, on the theory of the vaccinators, should have suffered exceptionally from small-pox; in the other we have a picked body of 220,000 men, who, on the evidence of the medical authorities, are as well protected as they know how to make them, and among whom, therefore, small-pox should be almost or quite absent, and small-pox deaths quite unknown. Let us see, then, what has happened in these two cases.

Perhaps the most remarkable and the most complete body of statistical evidence presented to the Commission was that of Mr. Thomas Biggs, a sanitary engineer and a town councillor of Leicester. It consists of fifty-one tables exhibiting the condition of the population in relation to health and disease from almost every conceivable point of view. The subject is further illustrated by sixteen diagrams, many of them in colours, calculated to exhibit to the eye in the most clear and simple manner the relations of vaccination and sanitation to small-pox and to the general health of the people, and especially of the children, in whose behalf it is always alleged vaccination is enforced. From this wealth of material I can give only two diagrams exhibiting the main facts of the case, as shown by Mr. Biggs' statistics in the Fourth Report of the Royal Commission, all obtained from official sources.

The first diagram (No. VIII.) shows in the upper part, by a dotted line, the total vaccinations, public and private, since 1850.[1] The middle line shows the mortality per million living from the chief zymotic diseases—fevers, measles, hooping-cough, and diphtheria — while the lower line gives the small-pox mortality. We notice here a high mortality from zymotics and from small-pox epidemics, during the whole period of nearly complete vaccination from 1854 to 1873. Then commenced the movement against vaccination, owing to its proved uselessness in the great epidemic, when Leicester had a very much higher small-pox mortality than London, which has resulted in a continuous decline, especially rapid for the last fifteen years, till it is now reduced to almost nothing. For that period, not only has small-pox mortality been continuously very low, but the zymotic diseases have also regularly declined to a lower amount than has ever been known before.

The second diagram (No. IX.) is even more important, as showing the influence of vaccination in increasing both the infantile and the total death-rates to an extent which even the strongest opponents of that operation had not thought possible. There are four solid lines on the diagram showing respectively, in five-year averages from 1838–42 to 1890–95, (1) the total death-rate per 1,000 living, (2) the infant death-rate under five years, (3) the same under one year, and (4), lowest of all, the small-pox death-rate under five years. The dotted line shows the percentage of total vaccinations to births.

[1] From 1850 to 1873 the private vaccinations have been estimated according to their proportion of the whole since they have been officially recorded.

The first thing to be noted is, the remarkable simultaneous rise of all four death-rates to a maximum in 1868–72, at the same time that the vaccination rate attained its maximum. The decline in the death-rates from 1852 to 1860 was due to sanitary improvements which had then commenced; but the rigid enforcement of vaccination checked the decline owing to its producing a great increase of mortality in children, an increase which ceased as soon as vaccination diminished. This clearly shows that the deaths which have only recently been acknowledged as due to vaccination, directly or indirectly, are really so numerous as largely to affect the total death-rate; but they were formerly wholly concealed, and are still partially concealed, by being registered under such headings as erysipelas, syphilis, diarrhœa, bronchitis, convulsions, or other proximate cause of death.[1]

Here, then, we have indications of a very terrible fact, the deaths, by various painful and often lingering diseases, of thousands of children as the result of that useless and dangerous operation termed vaccination. It is difficult to explain the coincidences exhibited by this diagram in any other way, and it is strikingly corroborated by a diagram of infant mortality in

[1] Mr. Biggs gave his evidence in 1891, and was obliged to rely on an estimate of the increase population since 1881. This was afterwards found to be too high; and the Commissioners urged that this would cause the decreased mortality during the decade to be greater than it really was. This is true; but the possible amount of the error is shown in the present diagrams by the added death-rates for 1893–96, which are calculated from the last census-populations. Thus, the only change produced in diagram IX. would be, that the decline from 1878–82 to 1893–96, would be a little more regular than is shown, while its general teaching would remain absolutely unaffected.

London and in England which I laid before the Royal Commission, and which I here reproduce (No. X.). The early part of this diagram is from a table calculated by Dr. Farr from all the materials available in the Bills of Mortality; and it shows for each twenty years the marvellous diminution in infant mortality during the hundred years from 1730 to 1830, proving that there was some continuous beneficial change in the conditions of life. The materials for a continuation of the diagram are not given by the Registrar-General in the case of London, and I have had to calculate them for England. But from 1840 to 1890 we find a very slight fall, both in the death-rate under five years and under one year for England, and under one year for London, although both are still far too high, as indicated by the fact that in St. Saviour's it is 213, and in Hampstead only 123 per 1,000 births. There appear to have been some causes which checked the diminution in London after 1840, then produced an actual rise from 1860 to 1870, followed by a slight but continuous fall since. The check to the diminution of the infant death-rate is sufficiently accounted for by that extremely rapid growth of London by immigration which followed the introduction of railways, and which would appreciably increase the child-population (by immigration of families) in proportion to the births. The rise from 1860 to 1870 exactly corresponds to the rise in Leicester, and to the strict enforcement of infant vaccination, which was continuously high during this period; while the steady fall since corresponds also to that continuous fall in the vaccination rate due to a growing conviction of its uselessness and its danger. These facts strongly support the contention that vaccination, instead of saving thousands of infant lives, as has been claimed, really destroys them by thousands, entirely neutralising that great reduction which was in progress from the last century, and which the general improvement in health would certainly have favoured. It may be admitted that the increasing employment of women in factories is also a contributory cause of infant mortality; but there is no proof that a less proportion of women have been thus employed during the last twenty years, while it is certain that there has been a great diminution of vaccination, which is now admitted to be a *vera causa* of infant mortality.

Before leaving the case of Leicester it will be instructive to compare it with some other towns of which statistics are available. And first, as to the great epidemic of 1871–2 in Leicester and in Birmingham. Both towns were then well vaccinated, and both suffered severely by the epidemic. Thus:

| | Leicester. | Birmingham. |
|---|---|---|
| S.P. cases per 10,000 population . . | 327 . | 213 |
| „ deaths „ „ „ . . | 35 . | 35 |

But since then, Leicester has rejected vaccination to such an extent that in 1894 it had only seven vaccinations to ten thousand population, while Birmingham had 240, or more than thirty times as much, and the proportion of its inhabitants who have been vaccinated is probably less than half those of Birmingham. The Commissioners themselves state that the disease was brought into the town of Leicester on twelve separate occasions during the recent epidemic, yet the following is the result:

| 1891–4. | Leicester. | Birmingham. |
|---|---|---|
| S.P. cases per 10,000 population . . | 19 . | 63 |
| „ deaths „ „ „ . . | 11 . | 5 |

Here we see, that Leicester had less than *one-third* the cases of small-pox, and less than *one-fourth* the deaths in proportion to population than well-vaccinated Birmingham; so that both the alleged *protection* from attacks of the disease, and *mitigation* of its severity when it does attack, are shown, not only to be absolutely untrue, but to apply really, in this case, to the absence of vaccination!

But we have yet another example of an extremely well-vaccinated town in this epidemic—Warrington, an official report on which has just been issued. It is stated that 99·2 per cent. of the population had been vaccinated, yet the comparison with unvaccinated Leicester stands as follows:

| EPIDEMIC OF 1892-3. | LEICESTER. | WARRINGTON. |
|---|---|---|
| S.P. cases per 10,000 population . . | 19·3 . | 123·3 |
| „ deaths „ „ „ . . | 1·4 . | 11·4 |

Here then we see that in the thoroughly vaccinated town the cases are more than six times, and the deaths more than eight times, that of the almost unvaccinated town, again proving that the most efficient vaccination does not *diminish* the number of attacks, and does not *mitigate* the severity of the disease, but that both these results follow from sanitation and isolation.

Now let us see how the Commissioners, in their *Final Report*, deal with the above facts, which are surely most vital to the very essence of the enquiry, and the statistics relating to which have been laid before them with a wealth of detail not equalled in any other case. Practically they ignore it altogether. Of course I am referring to the Majority Report, to which alone the Government and the unenlightened public are likely to pay any attention. Even the figures above quoted as to Leicester and Warrington are to be found only in the Report of the Minority, who also give the case of another town, Dewsbury, which has partially rejected vaccination, but not nearly to so large an extent as Leicester; and in the same epidemic it stood almost exactly between unvaccinated Leicester and well-vaccinated Warrington, thus:

| | | | | | | |
|---|---|---|---|---|---|---|
| Leicester . . . . | had | 1·1 | mortality | per | 10,000 | living |
| Dewsbury . . . . | „ | 6·7 | „ | „ | „ | „ |
| Warrington . . . | „ | 11·8 | „ | „ | „ | „ |

Here again we see that it is the unvaccinated towns that suffer least, not the most vaccinated. The public of course have been terrorised by the case of Gloucester, where a large default in vaccination was followed by a very severe epidemic of small-pox. The Majority Report refers to this in par. 373, intending to hold it up as a warning, but strangely enough in so important a document, say the reverse of what they mean to say, giving to it "very little," instead of "very much" small-pox. This case, however, has really nothing whatever to do with the question at issue, because, although anti-vaccinators maintain that vaccination has not the least effect in preventing or mitigating small-pox, they do *not* maintain that the *absence* of vaccination prevents it. What they urge is, that sanitation and isolation are the effective and only preventives; and it was because Leicester attended thoroughly to these matters, and Gloucester wholly neglected them, that the one suffered so little and the other so much in the recent epidemic. On this subject every enquirer should read the summary of the facts given in the Minority Report, paragraph 261.

To return to the Majority Report. Its references to Leicester are scattered over 80 pages, referring separately to the hospital staff, and the relations of vaccinated and unvaccinated to small-pox; while in only a few paragraphs (par. 480–486) do they deal with the main question and the results of the system of isolation adopted. These results they endeavour to minimise by declaring that the disease was remarkably "slight in its fatality," yet they end by admitting that "the experience of Leicester affords cogent evidence that the vigilant and prompt application of isolation . . . is a most powerful agent in limiting the spread of small-pox." A little further on (par. 500) they say, when discussing this very point—how far sanitation may be relied on in place of vaccination —"The experiment has never been tried." Surely a town of 180,000 inhabitants which has neglected vaccination for twenty years, is an experiment. But a little further on we see the reason of this refusal to consider Leicester a test experiment. Par. 502 begins thus: "The question we are now discussing must, of course, be argued on the hypothesis that vaccination affords protection against small-pox." What an amazing basis of argument for a Commission supposed to be enquiring into this very point! They then continue: "Who can possibly say that if the disease once entered a town the population of which was entirely or almost entirely unprotected, it would not spread with a rapidity of which we have in recent times had no experience?" But Leicester *is* such a town. Its infants—the class which always suffers in the largest numbers—are almost wholly unvaccinated, and the great majority of its adults have, according to the bulk of the medical supporters of vaccination, long outgrown the benefits, if any, of infant-vaccination. The disease has been introduced into the town twenty times before 1884, and twelve times during the last epidemic (*Final Report*, par. 482 and 483). The doctors have been asserting for years that once small-pox comes to Leicester it will run through the town like wild-fire. But instead of that it has been quelled with far less loss than in any of the best vaccinated towns in England. But the Commissioners ignore *this* actual experiment, and soar into the regions of conjecture with, "Who can possibly say?"—concluding the paragraph with—"*A priori* reasoning on such a question is of little or no value." Very true. But *a posteriori* reasoning, from the cases of Leicester, Birmingham, Warrington, Dewsbury, and Gloucester, *is* of value; but it is of value as showing the utter uselessness of vaccination, and it is therefore, perhaps, wise for the professional upholders of vaccination to ignore it. But surely it is *not* wise for a presumably impartial Commission to ignore it as it is ignored in this Report.[1]

[1] Although the Commission make no mention of Mr. Bigg's tables and diagrams showing the rise of infant-mortality with increased vaccination, and its fall as vaccination diminished, they occupied a whole day cross-examining him upon them, endeavouring by the minutest criticism to diminish their importance. Especially it was urged that the increase or decrease of mortality did not agree *in detail* with the increase or decrease of vaccination, forgetting that there are *numerous* causes contributing to all variations of death-rate, while vaccination is only alleged to be a *contributory* cause, clearly visible in general results, but not to be detected in smaller variations (see Fourth Report, Q. 17,513–17,744, or pp. 370 to 381). Mr. Bigg's cross-examination in all occupies 110 pages of the Report.

### *The Army and Navy as a Conclusive Test*

In the Report of the Medical Officer of the Local Government Board for 1884, it is alleged that, when an adult is revaccinated "he will receive the full measure of protection that vaccination is capable of giving him." In the same year the Medical Officer of the General Post Office stated in a circular, "It is desirable, in order to obtain full security, that the operation (vaccination) should be repeated at a later period of life"; and the circular of the National Health Society, already referred to, states that, "soldiers who have been revaccinated can live in cities intensely affected by small-pox without themselves suffering to any appreciable degree from the disease." Let us then see how far these official statements are true or false.

In their *Final Report* the Commissioners give the statistics of small-pox mortality in the Army and Navy from 1860 to 1894; and, although the latest order for the vaccination of the whole force in the Navy was only made in 1871, there can be no doubt that, practically, the whole of the men had been revaccinated long before that period;[1] but certainly since 1873 all without exception, both English and foreign, were revaccinated; and in the Army every recruit has been revaccinated since 1860 (see 2nd Report, Q. 3,453, 3,455; and for the Navy, Q. 2,645, 6, 3,212–13, and 3,226–3,229). Brigade-Surgeon William Nash, M.D., informed the Commission that the vaccination and revaccination of the Army was "as perfect as endeavours

[1] It was introduced into the Navy in 1801, and in that year the medical officers of the fleet presented Jenner with a special gold medal!

can make it," and that he can make no suggestion to increase its thoroughness (Q. 3,559, 3,560).

Turning now to the diagram (No. XI.) which represents the official statistics, the two lower lines show the small-pox death-rate per 100,000 of the force of the Army and Navy for each year, from 1860 to 1894. The lower thick line shows the Army mortality, the thin line that of the Navy. The two higher lines show the total death-rate from disease of the Navy, and of the Home force of the Army, as the tables supplied do not separate the deaths by disease of that portion of the Army stationed abroad.

Looking first at these upper lines, we notice two interesting facts. The first is, the large and steady improvement of both forces as regards health-conditions during the thirty-five years; and the second is the considerable and constant difference in the disease mortality of the two services, the soldiers having thoroughout the whole period a much higher mortality than the sailors. The decrease of the general mortality is clearly due to the great improvements that have been effected in diet, in ventilation, and in general health-conditions; while the difference in health between the two forces is almost certainly due to two causes, the most important being that the sailors spend the greater part of every day in the open-air, and in air of the maximum purity and health-giving properties, that of the open sea; while soldiers live mostly in camps or barracks, often in the vicinity of large towns, and in a more or less impure atmosphere. The other difference is, that soldiers are constantly subject to temptations and resulting disease, from which sailors while afloat are wholly free.

Turning now to the lower lines, we see that, as

regards small-pox mortality, the Navy suffered most down to 1880, but that since that period the Army has had rather the higher mortality. This has been held to be due to the less perfect vaccination of the Navy in the earlier period, but of that there is no proof, while there is evidence as to the causes of the improvement in general health. Staff-Surgeon T. J. Preston, R.N., stated them thus: "Shorter sea-voyages; greater care not to overcrowd; plentiful and frequent supplies of fresh food; the introduction of condensed water; and the care that is now taken in the general economy and hygiene of the vessels" (Q. 3,253). These seem sufficient to have produced also the comparative improvement in small-pox mortality, especially as the shorter voyages would enable the patients to be soon isolated on shore. The question we now have to consider is, whether the amount of small-pox here shown to exist in both Army and Navy demonstrates the "full security" that revaccination is alleged to give; whether as a matter of fact our soldiers and sailors, when exposed to the contagion of intense small-pox, *do* suffer to "any appreciable degree"; and lastly, whether they show any immunity whatever when compared with similar populations who have been either very partially or not at all revaccinated. It is not easy to find a fairly comparable population, but after due consideration it seems to me that Ireland will be the best available, as the statistics are given in the Commissioners' Reports, and it can hardly be contended that it has any special advantages over our soldiers and sailors,—rather the other way. I have therefore given a diagram, XII., in which a dotted line shows the small-pox mortality of the Irish people of the ages 15 to 45 in comparison with the Army and the

Navy mortality for the same years. (The figures for this diagram, as regards Ireland, have been calculated from the table at p. 37 of the *Final Report*, corrected for the ages 15 to 45 by means of Table J. at p. 274 of the Second Report.)

This dotted line shows us that, with the exception of the great epidemic of 1871, when for the bulk of the Irish patients there was neither isolation nor proper treatment, the small-pox mortality of the Irish population of similar ages has been on the average below that of either the Army or the Navy; while if we take the mean mortality of the three for the same period (1864–1894) inclusive, the result is as follows:

| | | | | | | |
|---|---|---|---|---|---|---|
| Army, mean of the annual | small-pox | death | rate, | 58 | per million. | |
| Navy " | " | " | " | 90 | " | |
| Ireland (ages 15–45) | " | " | " | 65·8 | " | [1] |

If we combine the Army and Navy death-rates in the proportion of their mean strength so as to get the true average of the two forces, the death-rate is **64·3** per million, or almost exactly the same as that of Ireland.

Now if there were no other evidence which gave similar results, this great test case of large populations

[1] These figures (for the Army and Navy) are obtained by averaging the annual death-rates given in the tables referred to, and are therefore not strictly accurate on account of the irregularly varying strength of the forces. But the error is small. In the case of the Navy, from 1864 to 1888 the tables enable the mortality to be accurately calculated, and the result comes out *more*, by nearly six per cent., than the mean above given; and in the case of the Army for the same years about one per cent. more. For Ireland the calculation has been accurately made by means of the yearly populations given at p. 37 of the *Final Report*, but for the Army and Navy materials for the whole period included in the diagrams are not available in any of the Reports.

compared over a long series of years, is alone almost conclusive; and we ask with amazement,—Why did not the Commissioners make some such comparison as this, and not allow the public to be deceived by the grossly misleading statements of the medical witnesses and official apologists for a huge imposture? For here we have, on one side a population which the official witnesses declare to be as well vaccinated and re-vaccinated as it is possible to make it, and which has all the protection that can be given by vaccination. It is a population which, we are officially assured, can live in the midst of the contagion of severe small-pox and not suffer from the disease "in any appreciable degree." And on comparing this population of over 200,000 men, thus thoroughly protected and medically cared for, with the poorest and least cared for portion of our country—a portion which the official witness regarding it declared to be *badly* vaccinated, while no amount of revaccination was even referred to—we find the less vaccinated and less cared for community to have actually a much lower small-pox mortality than the Navy, and the same as that of the two forces combined. The only possible objections that can be taken, or that were suggested during the examination of the witnesses are, that during the early portion of the period, the Navy was not *wholly* and *absolutely* revaccinated; and secondly, that troops abroad, and especially in India and Egypt, are more frequently subjected to infection. As to the first objection, even if revaccination were not absolutely universal in the Navy prior to 1873, it was certainly very largely practised, and should have produced a great difference when compared with Ireland. And the second objection is simply childish. For what are vaccination and revaccination for, except to protect from infection? And under exposure to the most intense infection they have been officially declared "not appreciably to suffer"!

But let us make one more comparison comprising the period since the great epidemic of 1871–2, during which the Navy as well as the Army are admitted to have been completely revaccinated, both English and foreign. We will compare this (supposed) completely protected force with Leicester, an English manufacturing town of nearly the same population, by no means especially healthy, and which has so neglected vaccination that it may now claim to be the least vaccinated town in the kingdom. The average annual small-pox death-rate of this town for the twenty-two years 1873–94 inclusive is thirteen per million (see 4th Report, p. 440); but in order to compare with our Army and Navy we must add one-ninth for the mortality at ages 15–45 as compared with total mortality, according to the table at p. 155 of the *Final Report*, bringing it to 14·4 per million, when the comparison will stand as follows:

| | Per Million. |
|---|---|
| Army (1873–94) small-pox death rate . . . | 37 [1] |
| Navy " " " . . . | 36·8 |
| Leicester " " ages 15–45 . . . | 14·4 |

[1] The figures for the Army are obtained from the Second Report, p. 278, down to 1888, the remaining six years being obtained from the *Final Report*, pp. 86, 87; but this small addition has involved a large amount of calculation, because the Commissioners have given the death-rates per 10,000 strength of four separate forces—Home, Colonial, Indian, and Egyptian, and have not given the figures for the whole Army, so as to complete the table in the Second Report. The figures for the Navy are obtained from the *Final Report*, p. 88.

It is thus completely demonstrated, that all the statements by which the public has been gulled for so many years, as to the almost complete immunity of the revaccinated Army and Navy, are absolutely false. It is all what Americans call "bluff." There is *no* immunity. They have *no* protection. When exposed to infection, they *do* suffer just as much as other populations, or even more. In the whole of the nineteen years 1878–1896 inclusive, unvaccinated Leicester had so few small-pox deaths that the Registrar-General represents the average by the decimal 0·01 per thousand population, equal to ten per million, while for the twelve years 1878–1889 there was less than one death per annum! Here we have *real* immunity, *real* protection; and it is obtained by attending to sanitation and isolation, coupled with the almost total neglect of vaccination. Neither Army nor Navy can show any such results as this. In the whole twenty-nine years tabulated in the Second Report the Army had not one year without a small-pox death, while the Navy never had more than three consecutive years without a death, and only six years in the whole period.

Now if ever there exists such a thing as a crucial test, this of the Army and Navy, as compared with Ireland, and especially with Leicester, affords such a test. The populations concerned are hundreds of thousands; the time extends to a generation; the statistical facts are clear and indisputable; while the case of the Army has been falsely alleged again and again to afford indisputable proof of the value of vaccination when performed on adults. It is important, therefore, to see how the Commissioners deal with these conclusive test-cases. They were appointed to discover the truth and to enlighten the public and the legislature, not merely to bring together huge masses of undigested facts.

What they do is, to make no comparison whatever with any other fairly comparable populations; to show no perception of the crucial test they have to deal with; but to give the Army and Navy statistics separately, and as regards the Army piecemeal, and to make a few incredibly weak and unenlightening remarks. Thus, in par. 333, they say that, during the later years, as the whole force became more completely revaccinated, small-pox mortality *declined.* But they knew well that during the same period it declined over all England, Scotland, and Ireland, with no special revaccination, and most of all in unvaccinated Leicester! Then with regard to the heavy small-pox mortality of the wholly revaccinated and protected troops in Egypt, they say, "We are not aware what is the explanation of this." And this is absolutely all they say about it! But they give a long paragraph to the Post Office officials, and make a great deal of *their* alleged immunity. But in this case the numbers are smaller, the periods are less, and no statistics whatever are furnished except for the last four years! All the rest is an extract from a parliamentary speech by Sir Charles Dilke in 1883, stating some facts, furnished of course by the medical officers of the Post Office, and therefore not to be accepted as evidence.[1] This slurring over the damning evidence of the

[1] Neither Sir C. Dilke nor the Post Office medical officers of the period referred to gave evidence before the Commission, and it shows to what lengths the Commissioners would go to support vaccination when such unverified verbal statements are accepted in their *Final Report.*

absolute inutility of the most thorough vaccination possible, afforded by the Army and Navy, is sufficient of itself to condemn the whole *Final Report* of the majority of the Commissioners. It proves that they were either unable or unwilling to analyse carefully the vast mass of evidence brought before them, to separate mere beliefs and opinions from facts, and to discriminate between the statistics which represented those great "masses of national experience" to which Sir John Simon himself has appealed for a final verdict, and those of a more partial kind, which may be vitiated by the prepossessions of those who registered the facts. That they have not done this, but without any careful examination or comparison have declared that revaccinated communities have "exceptional advantages" which, as a matter of fact, the Report itself show they have *not*, utterly discredits all their conclusions, and renders this *Final Report* not only valueless but misleading.

## V

### CRITICAL REMARKS ON THE "FINAL REPORT"

BEFORE proceeding to sum up the broad statistical case against vaccination, it may be well here to point out some of the misconceptions, erroneous statements, vague opinions, and conclusions which are opposed to the evidence, which abound in this feeble Report.

And first, we have the repetition of an oft-corrected and obviously erroneous statement as to the absolute identity of the vaccinated and the unvaccinated, except on the one point of vaccination. The Commissioners say: "Those, therefore, who are selected as being vaccinated persons might just as well be so many persons chosen at random out of the total number attacked.

So far as any connection with the incidence of, or the mortality from, small-pox is concerned, the choice of persons might as well have been made according to the colour of the clothes they wore (*Final Report*, par. 213). But there are tables in the Reports showing that about one-seventh of all small-pox deaths occur in the first six months of life, and by far the larger part of this mortality occurs in the first three months. The age of vaccination varies actually from three to twelve months, and many children have their vaccination specially delayed on account of ill-health, so that the "unvaccinated" always include a large proportion of those who, *merely because they are infants*, supply a much larger proportion of deaths from small-pox than at any other age. Yet the Commissioners say the unvaccinated might as well be chosen at random, or by the colour of their clothes so far as any liability to small-pox is concerned. One stands amazed at the hardihood of a responsible body of presumably sensible and truth-seeking men who can deliberately record as a fact what is so obviously untrue.

Hardly less important is it that the bulk of the unvaccinated, those who escape the vaccination officers, are the very poor, and the nomad population of the country—tramps, beggars and criminals, the occupants of the tenement houses and slums of our great cities, who, being all weekly tenants, are continually changing their residence. Such were referred to, in the Report of the Local Government Board for 1882 (p. 309), as constituting the bulk of the thirty-five thousand of default, under the heading—"Removed, not to be traced, or otherwise accounted for."

One of the Commission's official witnesses, Dr.

MacCabe, Medical Commissioner for Ireland, distinctly affirms this. He says (2nd Report, Q. 3,073) that he formerly had charge of the Dublin district, and that "out of a population of a quarter of a million, 100,000 live in tenement-houses, that is to say, houses that are let out in single rooms for the accommodation of a family. It is amongst that class, to a very great extent, that the defaulters exist. The relieving officer, when he goes to the tenement-dwelling where the birth occurred, finds that the parents have gone to some other tenement-dwelling and there is no trace of them . . . A great number of these defaulters occur in this way."

Now weekly tenants do not live in the best and most sanitary parts of towns, and the records of every epidemic show that such insanitary districts have an enormously greater proportion of the small-pox deaths than the healthier districts. Yet the Commissioners declare that there is "absolutely no difference between the vaccinated and the unvaccinated" except in respect of vaccination. Again we stand amazed at a statement so contrary to the fact. But the Commissioners must of course have believed it to be true, or they would not put it in their *Final Report*, upon which legislation may be founded affecting the liberties and the lives of their fellow countrymen.

I submit to my readers with confidence that this statement, so directly opposed to the clearest and simplest facts and to the evidence of official witnesses, proves the incapacity of the Commissioners for the important inquiry they have undertaken. By their treatment of this part of the subject they exhibit themselves as either ignorant or careless, in either case as thoroughly incompetent.

The next passage that calls for special notice here is par. 342, where they say, "We find that particular classes within the community, amongst whom revaccination has prevailed to an exceptional degree, have exhibited a position of quite exceptional advantage in relation to small-pox, although these classes have in many cases been subject to exceptional risk of contagion." It seems almost incredible that such a statement as this could be made as a conclusion from the official evidence before the Commissioners, and it can only be explained by the fact that they never made the simplest and most obvious comparisons, and that they laid more stress on bad statistics than on good ones. They trust, for example, to the cases of nurses in hospitals,[1] as to which there are absolutely no

[1] As regards the case of the nurses in small-pox hospitals, about which so much has been said, I brought before the Commission some evidence from a medical work, which sufficiently disposes of this part of the question. In Buck's *Treatise on Hygiene and the Public Health*, Vol. II., we find an article by Drs. Hamilton and Emmett on "Small-pox and other Contagious Diseases," and on page 321 thereof we read:

"It is a fact fully appreciated by medical men, that persons constantly exposed to small-pox very rarely contract the disease. In the case of physicians, health-inspectors, nurses, sisters of charity, hospital orderlies, and some others, this is the rule; and of over 100 persons who have been to my knowledge constantly exposed, some of them seeing as many as 1,000 cases, I have never personally known of more than one who has contracted the disease; but there are many writers who believe perfect immunity to be extremely rare. In this connection attention may be called to the exemption of certain persons who occupy the same room, and perhaps bed, with the patients, and though sometimes never vaccinated, altogether escape infection."

And Mr. Wheeler shows that at Sheffield the hospital staff *did* suffer from small-pox in a higher degree than other comparable populations (see 6th Report, Q. 19,907).

*statistics* in the proper sense of the term, only verbal statements by various medical men; and they overlook or forget the largest and only trustworthy body of statistics existing, as to revaccination—that of the Army and Navy! "A position of quite exceptional advantage!!" When the small-pox mortality of more than 200,000 men, all revaccinated to the completest extent possible by the medical officials, shows *no advantage whatever* over the whole comparable population of Ireland, and a quite exceptional *disadvantage* in comparison with almost unvaccinated Leicester![1] There is

[1] It is a common practice of vaccinists to quote the German Army as a striking proof of the good effects of revaccination; but as our own Army is as well vaccinated as the Army surgeons with unlimited power can make it, it is unlikely that the Germans can do so very much better. And there is some reason to think that *their* statistics are less reliable than our own. Lieut.-Col. A. T. Wintle, (late) R.A., has published in the *Vaccination Inquirer* extracts from a letter from Germany stating, on the authority of a German officer, that the Army statistics of small-pox are *utterly unreliable.* It is said to be the rule for Army surgeons to enter small-pox cases as skin-disease or some other "appropriate illness," while large numbers of small-pox deaths are entered as "sent away elsewhere." We had better therefore be content with our own Army and Navy statistics, though even here there is some concealment. In 1860 Mr. Duncombe, M.P., moved for a return of the disaster at Shorncliffe Camp, where, it was alleged, 30 recruits were vaccinated, and six died of the results, but the return was refused. A letter in the *Lancet* of July 7, 1860, from a "Military Surgeon" stated that numbers of soldiers have had their arms amputated in consequence of mortification after vaccination; and a Baptist minister and ex-soldier, the Rev. Frederick J. Harsant, gave evidence before the Commission of another Shorncliffe disaster in 1868, he himself, then a soldier, having never recovered, and having had unhealed sores on various parts of his body for more than 20 years. Eighteen out of the twenty men vaccinated at the same time suffered; some were months in hospital and in a much worse condition than himself (6th Report, p. 207). In

only one charitable explanation of such a "finding" as this—namely, that the Commissioners were by education and experience wholly incompetent to deal intelligently with those great masses of national statistics which alone can furnish conclusive evidence on this question.

At the end of the main inquiry, as to the effect of vaccination on small-pox (pp. 98, 99) the Commissioners adopt a very hesitating tone. They say that —"where vaccination has been most thorough the protection appears to have been greatest," and that "the revaccination of adults appears to place them in so favourable a condition as compared with the unvaccinated." But why say "appears" in both these cases? It is a question of fact, founded on ample statistics, which show us clearly and unmistakably—as in comparing Leicester with other towns—that vaccination gives *no protection whatever*, and that the best and most thorough revaccination, as in the Army and Navy, does *not protect at all!* It is no question of "appearing" to protect. As a fact, it does *not* protect, and does not *appear* to do so. The only explanation of the use of this word "appears" is, that the Commissioners have founded their conclusions, not upon the statistical evidence at all, but upon the *impressions* and *beliefs* of the various medical officials they examined, who almost all *assumed* the protection as an already established fact. Such was the case of the army-surgeon who declared that the deaths were much fewer than they would have been without revaccination; and who, on being asked why

the same volume is the evidence of twenty medical men, all of whom have witnessed serious effects produced by vaccination, some being of a most terrible and distressing character.

he believed so, answered that it was from reading of the small-pox mortality in pre-vaccination times! He had made no comparisons, and had no figures to adduce. It was his *opinion*, and that of the other medical officers, that it was so. And the Commissioners apparently had always held the same *opinions*, which, being confirmed by the opinions of other official witnesses, they concluded that comparisons of the revaccinated Army and Navy with ordinary death-rates were as unnecessary as they would certainly have been puzzling to them. Hence "appears" in place of "is" or "does"; and their seven conclusions as to the value and protectiveness of vaccination all under the heading—"We think," *not* "We are convinced," or "It has been proved to us," or "The statistics of the Army and Navy, of Ireland, of Leicester and of many other places, *demonstrate* the ("protectiveness" or "inutility"—as the case may be) of vaccination." I trust that I have now convinced my readers that the best evidence—the evidence to which Sir John Simon and Dr. Guy have appealed—DEMONSTRATES complete INUTILITY, as against what "appears" to the Commissioners and what they "think."

One other matter must be referred to before taking leave of the Commissioners. I have already shown how completely they ignore the elaborate and valuable evidence, statistical tables and diagrams, furnished by those who oppose vaccination, such as were brought before them by Mr. Biggs of Leicester, Mr. A. Wheeler, and Mr. William Tebb, who, though all were examined and cross-examined on the minutest details, might as well never have appeared so far as any notice in the Final Report is concerned. But there is also a very elaborate paper contributed by Dr. Adolf

Vogt, Professor of Hygiene and Sanitary Statistics in the University of Berne, who offered to come to London and submit to cross-examination upon it, which, however, the Commission did not consider necessary. This paper, a translation of which is printed in the Appendix to the 6th Report, p. 689, is especially valuable as the work of a thorough statistician, who, from his position, has access to the whole body of European official statistics, and his discussion goes to the very root of the whole question. The treatise is divided into nine chapters, and occupies thirty-four closely printed pages of the Blue Book; but, being an elaborate argument founded mainly on a scientific treatment of statistics, there was probably no member of the Commission capable of adequately dealing with it. Yet it is of more value than fully nine-tenths of the remainder of the voluminous reports, with their 31,398 questions and answers. Professor Vogt's treatise covers almost the whole ground, medical and statistical, and enforces many of the facts and arguments I have myself adduced. But there are two points which must be especially mentioned. His first chapter is headed—"*A Previous Attack of Small-pox does not Confer Immunity.*" I have long been of opinion that this was the case, and have by me a brief statement, written six years since, to show that the rarity of second attacks *may*, in all probability, be fully explained by the doctrine of chances. But I had not statistics sufficient to prove this. Professor Vogt, however, having the statistical tables of all Europe at his command, is able to show, not only that the calculus of probabilities itself explains the rarity of a second attack of small-pox, but that second attacks occur more frequently than they should do on the doctrine of chances alone, indicating that, instead of there being any *immunity*, there is really a somewhat increased *susceptibility* to a second attack![1] This being

[1] Brief statement of the argument:

The chances of a person having small-pox a second time may be roughly estimated thus: Suppose the average annual death-rate by small-pox to be 500 per million, and the average duration of life forty years. Then the proportion of the population that die of small-pox will be 500 × 40 = 20,000 per million. If the proportion of deaths to cases is one to five, there will be 100,000 cases of small-pox per million during the life of that million, so that one-tenth of the whole population will have small-pox once during their lives.

Now, according to the law of probabilities alone, the chances of a person having small-pox twice will be the square of this fraction, or *one-hundredth*: so that on the average only one person in 100 would have small-pox twice if it were a matter of pure chance, and if nothing interfered with that chance. But there are interferences which modify the result. (1) Those that die of the first attack cannot possibly have it a second time. (2) It is most frequent in the very young, so that the chances of having it later in life are not equal. (3) It is an especially *epidemic* disease, only occurring at considerable intervals, which reduces the chances of infection to those who have had it once. (4) It is probable that most persons are only liable to infection at certain periods of life, having passed which without infection they never take the disease. It seems probable, therefore, that these several conditions would greatly diminish the chances in the case of any person who had once had small-pox, so that perhaps, under the actual state of things, chance alone would only lead to one person in two hundred having the disease a second time.

The above is only an illustration of the principle. Professor Vogt goes more fully into the question, and arrives at the conclusion that out of every 1,000 cases of small-pox the *probability* is that ten will be second attacks. Then by getting together all the European observations as to the actual number of second attacks during various epidemics, the average is found to amount to sixteen in 1,000 cases, showing a considerable surplus beyond the number due to probability. Further, the proportion of deaths to attacks has from early times been observed to be

the case, it becomes really ludicrous to read the questions and answers, and the serious discussions, as to whether a "good vaccination" protects more or less than a previous attack of small-pox. Some think the protection is the same, but the greater number think it is not quite so much. Even the most ardent vaccinists do not claim a greater protection. But none of them ever doubt the *fact* of the protection gained by having had the disease, and yet none of them, nor any of the Commissioners, thought that any *evidence*, much less proof, of the fact itself was needed. They took it for granted. "Everybody knows it." "Very few people have small-pox a second time." No doubt. But very few people suffer from any special accident twice—a shipwreck, or railway or coach accident, or a house on fire; yet one of these accidents does not confer immunity against its happening a second time. The taking it for granted that second attacks of small-pox, or of any other zymotic disease, are of that degree of rarity as to *prove* some immunity or protection, indicates the incapacity of the medical mind for deal-

high for second attacks; and it has also been observed by many eminent physicians, whose statements are given, that second attacks are more common in the case of persons whose first attacks were very severe, which is exactly the reverse of what we should expect if the first attack really conferred any degree of immunity.

Now the whole theory of protection by vaccination rests upon the *assumption* that a previous attack of the disease is a protection; and Professor Vogt concludes his very interesting discussion by the remark: "All this justifies our maintaining that the theory of immunity by a previous attack of small-pox, whether the natural disease or produced artificially, must be relegated to the realm of fiction." If this is the case, the supposed *probability* or *reasonableness* of an analogous disease, vaccinia, producing immunity wholly vanishes.

ing with what is a purely statistical and mathematical question.

Quite in accordance with this influence of small-pox in rendering the patient somewhat more liable to catch the disease during any future epidemic, is the body of evidence adduced by Professor Vogt, showing that vaccination, especially when repeated once or several times, renders the persons so vaccinated *more* liable to take the disease, and thus actually increases the virulence of epidemics. This has been suspected by some anti-vaccinators; but it is, I believe, now for the first time supported by a considerable body of statistics.

The other important feature in Professor Vogt's memoir is the strong support he gives to the view that small-pox mortality is really—other things being approximately equal—a function of density of population. All the evidence I have adduced goes to show this, especially the enormously high small-pox death-rate in crowded cities in approximate proportion to the amount of crowding. Professor Vogt adds some remarkable statistics illustrating this point, especially a table in which the 627 registration districts of England and Wales are grouped according to their density of population, from one district having only sixty-four persons to a square mile to six which have 20,698 per square mile, another column showing in how many of the years during the period 1859–1882 there were any small-pox deaths in the districts. The result shown is very remarkable. In the most thinly populated district no small-pox death occurred in *any one* of the twenty-four years; in the most densely peopled districts small-pox deaths occurred in *every one* of the twenty-four years. And the frequency of the occurrence of small-pox in all the intervening groups of districts followed exactly the density of the population. Taking two groups with nearly the same population, the fourth group of 107 districts, with a total population of 1,840,581, had small-pox deaths in only five or six out of the twenty-four years in any of them; while the thirteenth group of thirteen districts, with a population of 1,908,838, had small-pox deaths in twenty-three out of the twenty-four years. But the first group had a density of 160 to the square mile, and the last had 8,350 to the square mile. The Commissioners dwell upon the alleged fact that neither water-supply, nor drainage, nor contaminated food produce small-pox, and urge that what is commonly understood by sanitation has little effect upon it (par. 153). But what may be termed the fundamental principle of sanitation is the avoidance of *overcrowding*; and this is shown by an overwhelming body of evidence invariably to influence small-pox mortality quite irrespective of vaccination.[1] Yet the remarkable contribution to the mass of evidence in the "Reports" which brings out this fact most clearly, receives no notice whatever in the *Final Report*.

[1] It is not alleged that over-crowding, *per se*, is the direct *cause* of small-pox, or of any other zymotic disease. It is, perhaps rather a condition than a cause; but under our present social economy it is so universally associated with various causes of disease—impure air, bad drainage, bad water supply unhealthy situations, unwholesome food, overwork, and filth of every description in houses, clothing, and persons—that it affords the most general and convenient indication of an unhealthy as opposed to a healthy mode of life; and, while especially apply, ing to zymotic diseases, is also so generally prejudicial to health as to produce a constant and very large effect upon the total mortality.

## VI

### SUMMARY AND CONCLUSION

As the diverse aspects of the problem which has been discussed in the preceding pages are somewhat numerous and complex, owing to the vast mass of irrelevant but confusing matter with which it has been encumbered at every step of its progress for nearly a century, a brief summary of the main points here referred to, and a statement of their bearing on the essential problem, will now be given.

I have first shown the nature of the tests which seemed to the early enquirers to establish the protective influence of vaccination, and have given the facts which the two greatest living specialists on the subject—Professor Crookshank and Dr. Creighton—consider to prove the fallacy or insufficiency of all the tests which were applied. This is followed by a statement of the abundant evidence which, in the first ten years of the century, already showed that vaccination had no protective power (pp. 219–21). But the heads of the medical profession had accepted the operation as of proved value, and the legislature, on their recommendation, had voted its discoverer £30,000 of public money, and had besides, in 1808, endowed a National Vaccine Establishment with about £3,000 a year. Reputations and vested interests were henceforth at stake, and those who adduced evidence of the failure or the dangers of vaccination were treated as fanatics, and have been so treated by the medical and official world down to the appointment of the last Royal Commission.

I next give the reasons why doctors are *not* the best judges of the effects, beneficial or otherwise, of vaccination, and follow this by proofs of a special capacity for mis-stating facts in reference to this question which has characterized them from the beginning of the century down to our day. The successive annual reports of the National Vaccine Establishment give figures of the deaths by small-pox in London in the eighteenth century, which go on increasing like Falstaff's men in buckram; while in our own time the late Dr. W. B. Carpenter, Mr. Ernest Hart, the National Health Society, and the Local Government Board make statements or give figures which are absurdly and demonstrably incorrect (pp. 223–28).[1]

I then show the existence of so unreasoning a belief

[1] To the cases I have already given I may now add two others, because they illustrate the recklessness in making assertions in favour of vaccination which scorns the slightest attempt at verification. In the first edition of Mr. Ernest Hart's *Truth about Vaccination* (p. 4), it is stated, on the authority of a member of Parliament recently returned from Brazil, that during an epidemic of small-pox at the town of Ceara in 1878 and 1879, out of a population not exceeding 70,000 persons there were 40,000 deaths from small-pox. This was repeated by Dr. Carpenter during a debate in London, in February, 1882, and only when its accuracy was called in question was it ascertained that at the time referred to the population of Ceara was only about 20,000; yet the M.P. had stated—with detailed circumstance—that "in one cemetery, from August 1878, to June 1879 27,064 persons who had died of small-pox had been buried." Gazetteers are not very recondite works, and it would have been not difficult to test some portion of this monstrous statement before printing it. Jenner's biographer tells us that he had a horror of arithmetical calculations, due to a natural incapacity, which quality appears to be a special characteristic of those who advocate vaccination, as the examples I have given sufficiently prove.

Another glaring case of official misrepresentation occurred in the Royal Commission itself, but was fortunately exposed later on. A medical officer of the Local Government Board gave evidence

in the importance of vaccination that it leads many of those who have to deal with it officially to conceal-

(First Report, Q. 994) that the Board in 1886 "took some pains to get the figures as to the steamship *Preussen*," on which small-pox broke out on its arrival in Australia. He made the following statements: (1) There were 312 persons on board this vessel. (2) 4 revaccinated, 47 vaccinated, 3 who had small-pox, and 15 unvaccinated were attacked—69 in all. (3) The case was adduced to show that "sanitary circumstances have little or no control over small-pox compared with the condition of vaccination or no vaccination."

This official statement was quoted in the House of Commons as strikingly showing the value of vaccination. But, like so many other official statements, it was all wrong! The reports of the Melbourne and Sydney inspectors have been obtained, and it is found: (1) That there were on board this ship 723 passengers and 120 crew—843 in all, instead of 312; so that the "pains" taken by the Local Government Board to get "the figures" were very ineffectual. (2) There were 29 cases among the 235 passengers who disembarked at Melbourne, of whom only 1 was unvaccinated. The crew had *all* been revaccinated before starting, yet 14 of them were attacked, and one died. All these in addition to the cases given by the Local Government Board. Thus 18 revaccinated persons caught the disease, instead of 4, as first stated, and 69 vaccinated, instead of 48; while among the 15 cases alleged to be unvaccinated *three* were infants under one year old, and *two* more between five and ten years. (3) The official reports from Melbourne and Sydney stated that the vessel was greatly overcrowded, that the sanitary arrangements were very bad, and the inspector at Sydney declared the vessel to be the "filthiest ship he had had to deal with"! (See *Final Report*, pp. 205-6; and Second Report, Q. 5,942-5,984.)

Here, then, we have a case in which *all* the official *figures*, paraded as being the result of "taking some pains," are wrong, not to a trifling extent, but so grossly that they might be supposed to apply to some quite different ship. And the essential fact of the filthy, overcrowded, and unsanitary condition of the ship was unknown or concealed; and the case was adduced as one showing how unimportant is sanitation as regards small-pox. What the case really proves is, that under unsanitary conditions neither vaccination nor revaccination has the

ments and misstatements which are justified by the desire to "save vaccination from reproach." Thus it happened that till 1881 no deaths were regularly recorded as due to vaccination, although an increasing number of such deaths now appear in the Registrar-General's Reports; while a few medical men, who have personally inquired into these results of vaccination, have found a large amount of mortality directly following the operation, together with a large percentage of subsequent disease, often lasting for years or during life, which, except for such private enquiries, would have remained altogether unknown and unacknowledged (pp. 228–32).[1]

slightest effect in preventing the spread of small-pox, since the proportion of the cases among the revaccinated crew was almost exactly the same as that of the whole of the cases (omitting the three infants) to the whole population on the ship.

With this example of officially quoted *facts* (!) in support of vaccination, coming at the end of the long series we have given or referred to in the first part of this work, it is not too much to ask that all such unverified statements be, once and for ever, ruled out of court.

[1] The Commissioners in their Majority Report (par. 379) suggest that the deaths due to vaccination are small in proportion to the whole number of vaccinations, and argue that it would be as unwise to reject vaccination for this reason as to refuse to travel by railway on account of the risk of accident. But they overlook the fact that railway-travelling is not made compulsory; and they make no comparison by figures, showing the proportionate risk in the two cases. This comparison is however made in the Minority Report (par. 184), and it is shown, that the risk of death by vaccination, as officially admitted, is, on the average of the nine years 1881–89, between *three and four thousand times* greater than the risk of death by railway-travelling, while it is 622 times as great as in the very worst year of railway accidents, 1889. Supposing railway accidents and resulting deaths were 3,000 times as numerous as they are, should we be satisfied with the railway-companies'

The same desire to do credit to the practice which they believe to be so important, leads to such imperfect or erroneous statements as to the *vaccinated* or *unvaccinated* condition of those who die of small-pox, as to render all statistics of this kind faulty and erroneous to so serious an extent that they must be altogether rejected. Whether a person dies of small-pox or of some other illness is a fact that is recorded with tolerable accuracy, because the disease, in fatal cases, is among the most easily recognised. Statistics of "small-pox mortality" may, therefore, be accepted as reliable. But whether the patient is registered as vaccinated or not vaccinated usually depends on the visibility or non-visibility of vaccination-marks, either during the illness or after death, both of which observations are liable to error, while the latter entails a risk of infection which would justifiably lead to its omission. And the admitted practice of many doctors, to give vaccination the benefit of any doubt, entirely vitiates all such statistics, except in those special cases where large bodies of adults are systematically vaccinated or revaccinated. Hence, whenever the results of these *imperfect* statistics are opposed to those of the official records of small-pox *mortality*, the former must be rejected. It is an absolute law of evidence, of statistics, and of common sense, that when two kinds of evidence contradict each other, that which can be proved to be even partially incorrect or untrustworthy must be rejected. It will be found that *all* the evidence that seems to prove the value of

assurance, that it was really of no importance as compared with the benefits of railways! And the actual deaths from vaccination are, certainly, much greater than the officially admitted deaths used in the above calculation.

vaccination is of this untrustworthy character. This conclusion is enforced by the fact that the more recent hospital statistics show that small-pox occurs among the vaccinated in about the same proportion as the vaccinated bear to the whole population; thus again indicating that the earlier figures, showing that they were proportionately five or six times as numerous, and the death-rate of the unvaccinated twice or thrice that of the average of pre-vaccination days, are altogether erroneous, and are due to the various kinds of error or misstatement which have been pointed out (pp. 229–32).

Having thus cleared away some of the misconceptions and fallacies which have obscured the main question at issue, and having shown that, by official admission, the only valuable evidence consists of "large masses of national statistics," which should have been dealt with by a commission of trained statisticians, I proceed to show, by a series of diagrams embodying the official or national statistics brought before the Commission, or to be found in the Reports of the Registrar-General, what such statistics really prove; and I ask my readers to look again at those diagrams as I refer to them.

Diagram I. exhibits the most extensive body of national statistics available, showing at one view the death-rates from Small-pox, from the other chief Zymotic Diseases, and the Total Mortality, from 1760 to 1896. The first portion, from 1760 to 1836, is from the "Bills of Mortality," which, though not complete, are admitted to be, on the whole, fairly accurate as regards the variations at different periods and between different diseases. The second part, from 1838 onwards, is from the Reports of the Registrar-General, and is more complete in giving all deaths whatever. Its lines are, therefore, as it were, on a higher level than those of the earlier period, and can only be compared with it as regards proportions of the different mortalities, not so accurately as to their total amounts. The main teaching of this diagram—a teaching which the Commissioners have altogether missed by never referring to diagrams showing comparative mortalities —is the striking correspondence in average rise and fall of the death-rates of small-pox, of zymotics, and of all diseases together. This correspondence is maintained throughout the whole of the first part, as well as through the whole of the second part, of the diagram; and it proves that small-pox obeys, and always has obeyed, the same law of subservience to general sanitary conditions as the other great groups of allied diseases and the general mortality. Looking at this most instructive diagram, we see at once the absurdity of the claim that the diminution of small-pox in the first quarter of our century was due to the partial and imperfect vaccination of that period. Equally absurd is the allegation that its stationary character from 1842 to 1872, culminating in a huge epidemic, was due to the vaccination then prevailing, though much larger than ever before, not being quite universal—an allegation completely disproved by the fact that the other zymotics as a whole, as well as the general mortality, exhibited strikingly similar decreases followed by equally marked periods of average uniformity or slight increase, to be again followed by a marked decrease. There is here no indication whatever of vaccination having produced the slightest effect on small-pox mortality.

The second diagram shows that, even taking the Commission's favourite method of comparing the zymotics separately with small-pox, all of them except measles show a similar or a greater decrease during the period of official registration, and also agree in the periods of slight increase, again proving the action of the same general causes (which I have pointed out at p. 250), and leaving no room whatever for the supposed effects of vaccination.

Diagram III. shows that similar phenomena occurred in England and Wales as a whole, the other zymotics and the total deaths obeying the same laws of increase and decrease as small-pox. Comparison with diagram I. shows the much greater severity of small-pox epidemics in London, illustrating the fact, which all the statistical evidence of all countries strikingly enforces, that *small-pox mortality is, other things being equal, a function of density of population*, while it pays no regard whatever to vaccination. This is further shown by the short, thick dotted line which exhibits the total number of vaccinations since 1872, when private as well as public vaccinations were first officially recorded, and which proves that the continuous *decrease* of vaccination since 1882 has been accompanied by a decided *decrease*, instead of an increase, in small-pox mortality.

Diagram IV. shows the statistics of mortality in Ireland and Scotland from small-pox and certain chosen zymotics, from the tables which were laid before the Commission by the official advocates of vaccination. These show two striking facts, which the Commissioners failed to notice in their *Final Report*. First, the smaller amount of small-pox mortality in Ireland than in Scotland, the latter being alleged to be well vaccinated, the former imperfectly so; and, secondly, the similar difference in the two chosen diseases and the general parallelism of the two. Here again we see clearly the influence of *density of population*, Scotland having a very much larger proportion of its inhabitants living in large manufacturing towns.

The next three diagrams, V., VI., and VII., show small-pox mortality in Sweden, Prussia, and Bavaria—countries which at previous enquiries were adduced as striking examples of the value of vaccination. They all show phenomena of the same character as our own country, but far worse as regards epidemics in the capitals; that of Stockholm, in 1874, causing a death-rate more than 50 per cent. higher than during the worst epidemic of the last century in London! The diagram of small-pox and zymotics in Bavaria is given merely because the statistics were brought before the Commission as a proof of the beneficial results of vaccination in well-vaccinated communities. It was *alleged* by Dr. Hopkirk that almost the whole of the population were vaccinated, and *admitted* by him that of the 30,742 cases of small-pox in 1871 no less than 95·7 per cent. were vaccinated! The epidemic was, however, less severe than in Prussia, again showing the influence of density of population, less than one-seventh of the Bavarians inhabiting towns of over 20,000, while one-fourth inhabit similar towns in Prussia; but we see that during the latter half of the period chosen small-pox greatly increased, and the other zymotics remained very high, indicating general insanitary conditions. And this case was specially brought before the Commission as a proof of the *benefits* of vaccination! In their *Final Report* the

Commissioners omit to point out that it really indicates the very reverse.

We then come to the two cases that afford most conclusive tests of the absolute uselessness of vaccination—Leicester, and our Army and Navy.

Diagram VIII. shows the death-rates from small-pox and from the other zymotics in LEICESTER during the period of official registration, together with the percentage of vaccinations to births. Up to 1872 Leicester was a fairly well-vaccinated town; yet for thirty-four years its small-pox mortality, in periodical epidemics, remained very high, corresponding generally with the other zymotics. But immediately after the great epidemic of 1872, which was much worse than in London, the people began to reject vaccination, at first slowly, then more rapidly, till for the last eight years less than 5 per cent. of the births have been vaccinated. During the whole of the last twenty-four years small-pox deaths have been very few, and during twelve consecutive years, 1878–89, there was a total of only eleven small-pox deaths in this populous town.

Diagram IX. is equally important as showing a remarkable correspondence, if not a causal relation, between vaccination and disease. From 1848 to 1862 there was a considerable decrease of both general and infant mortality, and also in infant mortality from small-pox. This, Mr. Biggs tells us, was when important sanitary improvements were in progress. Then, the more thorough enforcement of vaccination set in (as shown by the dotted line), and was accompanied by an increase of all these mortalities. But, so soon as the revolt against vaccination began, till the present time when it has diminished to about 2 or 3 per cent. of births, all mortalities have steadily decreased, and that decrease has been especially marked in infant lives. It is very suggestive that the lines of infant mortality have now reached the position they would have had if the slow decrease during 1850–60 had been continued, strongly indicating that some *special cause* sent them up, and the *removal of that cause* allowed them to sink again; and during that very period *vaccination* increased and then steadily decreased. I venture to declare that in the whole history of vaccination there is no such clear and satisfactory proof of its having saved a single life, as these Leicester statistics afford of its having been the cause of death to many hundreds of infants.

Diagram X. exhibits the check to the decrease in infant mortality, both in London and for England, since the enforcement of vaccination (p. 57), and thus supports and enforces the conclusions derived from the preceding diagram.

I next discuss in some detail what is undoubtedly the most complete and crucial test of the value or uselessness of vaccination to be found anywhere in the world. Since 1860 in the Army, and 1872 in the Navy, every man without exception, English or foreign, has been vaccinated on entering the service, though for long before that period practically the whole force was vaccinated or revaccinated. Diagrams XI. and XII. exhibit the result of the statistics presented to the Commission, showing for the Navy, the death-rate from disease and that from small-pox for the whole force; and for the Army, the death-rate from small-pox for the whole force, and that from disease for the home force only, foreign deaths from disease not being separately given.

Here we note, first, the general parallelism of the two lines showing the diminishing total disease-mortality in the two forces, resulting from the greater attention given to sanitation and to general health conditions of both forces during the last thirty or forty years. But, instead of small-pox mortality, as shown by the two lower lines of the diagram, absolutely vanishing with the complete revaccination in the Army since 1860, it shows but a small improvement as compared with general disease-mortality; just as if some adverse cause were preventing the improvement. In the Navy the improvement is somewhat greater, and more nearly comparable with that of general disease mortality. There is, therefore, as regards proportionate decrease, no indication whatever of any exceptional cause favourably influencing small-pox.

In diagram XII. I compare the small-pox mortality of the Army and Navy with that of Ireland, from tables given in the *Final Report* and the Second Report; and we find that this whole country (at ages 15–45) has actually a much lower small-pox mortality than the Army, while it is a little more than in the Navy, although the mortality during the great epidemic was higher than any that affected the Army or Navy, owing to its rapid spread by infection in the towns. But the proportionate numbers dying of small-pox in a series of years is, of course, the final and absolute test; and, applying this test, we find that these revaccinated soldiers and sailors have suffered in the thirty-one years during which the materials for comparison exist, to almost exactly the same extent as poor, half-starved, imperfectly vaccinated Ireland (p. 282)! Another and still more striking comparison is given. The town of Leicester is, and has been for the last twenty years, the least vaccinated town in the kingdom. Its average population from 1873 to 1894 was about two-thirds that of the Army during the same period. Yet the small-pox deaths in the Army and Navy were thirty-seven per million, those of Leicester under fifteen per million.

Thus, whether we compare the revaccinated and thoroughly "protected" Army and Navy with imperfectly vaccinated Ireland, or with almost unvaccinated Leicester, we find them either on a bare equality or *worse off* as regards small-pox mortality. It is not possible to have a more complete or crucial test than this is, and it absolutely demonstrates the utter uselessness, or worse than uselessness, of revaccination![1]

In the face of this clear and indisputable evidence, all recorded in their own Reports, the Commissioners make the astounding statement: "We find that particular classes within the community amongst whom revaccination has prevailed to an exceptional degree have exhibited a position of quite exceptional advantage in relation to small-pox, although these classes have in many cases been subject to exceptional risk of contagion" (*Final Report*, p. 90, par. 342). And again: "The fact that revaccination of adults appears to place them in so favourable a condition as compared with the unvaccinated," etc. (*Final Report*, p. 98, par. 375). What can be said of such statements as these, but simply that they are wholly untrue. And the

[1] So late as 1892 (Jan. 16) the *Lancet* declared in a leading article: "No one need die of small-pox; indeed, no one need have it unless he likes—that is to say, he can be absolutely protected by vaccination once repeated." Surely, never before was misstatement so ignorantly promulgated, or so completely refuted!

fact that the majority of the Commissioners did not know this, because they never compared the different groups of facts in their own reports which prove them to be untrue, demonstrates at once their complete incapacity to conduct such an inquiry and the utter worthlessness of their *Final Report.*

This is a matter upon which it is necessary to speak plainly. For refusing to allow their children's health, or even their lives, to be endangered by the inoculation into their system of disease-produced matter, miscalled "lymph,"[1] hundreds and probably thousands of English parents have been fined or imprisoned and treated as criminals; while certainly thousands of infants have been officially done to death, and other thousands injured for life. And all these horrors on account of what Dr. Creighton has well termed a "grotesque superstition," which has never had a rational foundation either of physiological doctrine or of carefully tested observations, and is now found to be disproved by a century's dearly bought experience. This disgrace of our much-vaunted scientific age has been throughout supported by concealment of facts telling against it, by misrepresentation, and by untruths. And now a Royal Commission, which one would have supposed would have striven to be rigidly impartial, has presented a Report which is not only weak, misleading, and inadequate, but is also palpably one-sided, in that it omits in every case to make those comparisons by which alone the true

[1] "LYMPH, a colourless nutritive fluid in animal bodies" (Chambers' Dictionary). How misleading to apply this term to a product of *disease*, used to produce another *disease*, and now admitted to be capable of transmitting some of the most *horrible diseases* which afflict mankind—syphilis and leprosy!

meaning can be ascertained of those "great masses of national experience" to which appeal has been made by the official advocate of vaccination *par excellence*—Sir John Simon.

I venture to think that I have here so presented the best of these statistical facts, as to satisfy my readers of the certain and absolute *uselessness* of vaccination as a preventive of small-pox; while these same facts render it in the highest degree probable that it has actually increased susceptibility to the disease. The teaching of the whole of the evidence is in one direction. Whether we examine the long-continued records of London mortality, or those of modern registration for England, Scotland, and Ireland; whether we consider the "control experiment" or crucial test afforded by unvaccinated Leicester, or the still more rigid test in the other direction, of the absolutely revaccinated Army and Navy, the conclusion is in every case the same: that vaccination is a gigantic delusion; that it has never saved a single life; but that it has been the cause of so much disease, so many deaths, such a vast amount of utterly needless and altogether undeserved suffering, that it will be classed by the coming generation among the greatest errors of an ignorant and prejudiced age, and its penal enforcement the foulest blot on the generally beneficent course of legislation during our century.

To talk of amending such legislation is a mockery. Absolute and immediate abolition is the only rational course open to us. Every day the vaccination laws remain in force, parents are being punished, infants are being killed. An Act of a single clause will repeal these vile laws; and I call upon every one of our legislators to consider their responsibilities as the guardians of the liberties of the English people, and to insist that this repeal be effected without a day's unnecessary delay.

The successive Vaccination Acts were passed by means of allegations which were wholly untrue and promises which have all been unfulfilled. They stand alone in modern legislation as a gross interference with personal liberty and the sanctity of the home; while as an attempt to cheat outraged nature and to avoid a zymotic disease without getting rid of the foul conditions that produce or propagate it, the practice of vaccination is utterly opposed to the whole teaching of sanitary science, and is one of those terrible blunders which, in their far-reaching evil consequences, are worse than the greatest of crimes.

# APPENDIX.

## THE CAUSES OF THE IMPROVEMENT IN THE HEALTH OF LONDON TOWARDS THE END OF THE EIGHTEENTH AND BEGINNING OF THE NINETEENTH CENTURIES.

ALTHOUGH, as I have shown, there is ample proof of the great improvement in the sanitary condition of London during the latter part of the eighteenth and the early part of the nineteenth centuries in the great and exceptional decrease of the general death-rate, and especially in the infant death-rate as pointed out by the late Dr. Farr, it will be well to give a brief sketch of the various changes, not only in London itself, but in the habits, and especially in the food of the people, which combined to bring it about.

In the early part of the eighteenth century London was in a condition of overcrowding and general filth which we can now hardly realize. The houses were low and overhung the streets, and almost all had cesspools close behind or underneath them. The streets were narrow, the main thoroughfares only being paved with cobble stones, which collected filth, and allowed it to soak into the ground beneath till the soil and subsoil became saturated. Slops and refuse of all kinds were thrown into the streets at night, and only the larger streets were ever cleaned. The bye-streets and the roads outside London were so bad that vehicles could only go two or three miles an hour; while even between London and Kensington, coaches sometimes stuck in the mud or had to turn back and give up the journey. The writers of the time describe the streets as dangerous and often impassable, while only in the main thoroughfare were there any footways, which were separated from the narrow roadway by rows of posts. Gay, in his *Trivia*, speaks of the slops thrown from the overhanging windows, and the frequent dangers of the night, adding,—

> "Though expedition bids, yet never stray
> Where no rang'd posts defend the rugged way."

And throughout his poem, dirt, mire, mud, slime, are continually referred to as being the chief characteristics of the streets. They mostly had a gutter on each side, and with few exceptions rain alone prevented their being blocked with refuse. The effects of a heavy shower in the city are forcibly described by Swift in his usual plain language,—

> "Now from all parts the swelling kennels flow,
> And bear their trophies with them as they go;
> Filths of all hues and odours seem to tell
> What street they sailed from by their sight and smell.
> * * * * *
> Sweepings from butchers' stalls, dung, guts, and blood,
> Drown'd puppies, stinking sprats, all drench'd in mud,
> Dead cats, and turnip tops, come tumbling down the flood."

Macaulay tells us that down to 1726, St. James' Square, though surrounded by houses of the nobility, was a common receptacle for refuse of all kinds, and that it required an Act of Parliament to stop its being so used. Hogs were kept in St. George's, Hanover Square, and in 1760 many were seized as a common nuisance.

The numerous small streams which flowed through London from the northern heights—Langbourne, Wallbrook, Fleet, Tybourne, and Westbourne—which were in earlier times a source of health and water-supply, gradually became noisome open sewers, and one after another were arched over. There were many wells in London, indicated by such names as Holywell, Clerkenwell, and Aldgate Pump, and there were also conduits in Cheapside and Cornhill; but it is certain that, from the filthy streets and house-cesspools, all the water derived from them must have been contaminated, and thus helped to produce the terrible mortality from plague and fevers of the seventeenth century. It has been often suggested that the Great Fire of London in 1666 was the cause of the final disappearance of the plague, but how, except that the new houses were for once clean and wholesome, has not, I think, been satisfactorily explained. I believe, however, that it can be found in the action of the fire upon the soil, which for more than a thousand years had been continuously saturated with filth, and must, as we now know, have afforded a nidus for every kind of disease-germs. The long-continued fire not only destroyed the closely-packed houses, but in doing so must have actually burnt the whole soil to a considerable depth, and thus have destroyed not only the living germs, but all the organic matter in it. The new city, for the first time for many centuries, had beneath it a dry and wholesome soil, which to this day has not had time to get so foully polluted as before the fire.

When we come to consider how the people lived, the conditions were equally bad. The houses were often sunk below the level of the ground and had very low rooms, as indicated by Gay's lines on the Strand,—

> "Where the low penthouse bows the walker's head,
> And the rough pavement wounds the yielding tread."

Light and air were shut out by the overhanging of each successive floor, and by enormous signboards projecting over the street; while any effective ventilation was out of the question, and, indeed, was never thought of. Water had usually to be brought from the public wells or conduits, and was used sparingly; and most business people lived for whole days and weeks without ever leaving the polluted air of their shops and houses. A friend of Mr. William White told him that he served his apprenticeship to a grocer in Cheapside from 1786 to 1793; that the shop was opened at seven in the morning and closed at ten at night; that he slept under the counter; that his ablutions were limited to his face and hands, and that he never went out except to meeting on Sunday. Bishop Wilson of Calcutta was in a silk-merchant's shop about the same time, and worked similar hours. He records, that the apprentices rarely left the house for weeks together, and that it was three years before he had his first holiday. William Cobbett, in 1783, was in a lawyer's office in Gray's Inn, where, he relates, "I worked like a galley slave from five in the morning till eight or nine at night, and sometimes all night long. I never quitted this gloomy recess except on Sundays, when I usually took a walk to St. James' Park." [1]

When we remember the filthy condition of the streets, and that, owing to the cesspools either under or close behind the houses, the scarcity of water, and the absence of ventilation, the shops and living rooms were always full of foul air, bad smells, and poisonous gases, how can we wonder at the prevalence of zymotic disease and the dreadful amount of infant and general mortality? And in many houses there was an additional peril in the vicinity of churchyards. In Nicholl's *Illustrations*

[1] White's *Story of a Great Delusion*, p. 81.

*of Literary History* (vol. iv. p. 499), Mr. Samuel Gale is quoted as writing, in 1736, as follows:—

"In the churchyard of St. Paul, Covent Garden, the burials are so frequent that the place is not capacious enough to contain decently the crowds of dead, some of whom are not laid above a foot under the loose earth. The cemetery is surrounded every way with close buildings; and an acquaintance of mine, whose apartments look into the churchyard, hath averred to me, that the family have often rose in the night time and been forced to burn frankincense and other perfumes to dissipate and break the contagious vapour. This is an instance of the danger of infection proceeding from the corrupt effluvia of dead bodies."

Many illnesses then originated in churches, and even those whose houses were exceptionally wholesome were often exposed to a dangerous atmosphere when they went to church on Sundays.

The general food of the poor and the middle classes added greatly to their unhealthiness, and itself caused disease. Owing to the absence of good roads, it was impossible to supply the large population of London with fresh food throughout the year, and, consequently, salt meat and salt fish formed the staple diet during the winter. For the same reason fresh vegetables were unattainable; so that meat, cheese, and bread, with beer as the common drink at all meals, was the regular food, with chiefly salted meat and fish in winter. As a result, scurvy was very common. Dr. Cheyne, in 1724, says, "There is no chronical distemper more universal, more obstinate, and more fatal in Britain, than the Scurvy." And it continued to be common down to 1783, when Dr. Buchan says, "The disease most common in this country is the Scurvy." But very soon afterwards it decreased, owing to the growing use of potatoes and tea, and an increased supply of fresh vegetables, fruit, milk, etc., which the improved roads allowed to be brought in quantities from the surrounding country.

Now it is quite certain, that the excessively unhealthy conditions of life, as here briefly described, continued with very partial amelioration throughout the middle portion of the century; and we have to consider what were the causes which then came into operation, leading to the great improvement in health that undoubtedly occurred in the latter portions of it and in the early part of our century.

Beginning with improvements in the streets and houses, we

have, in 1762, an Act passed for the removal of the overhanging signboards, projecting waterspouts, and other such obstructions. In 1766 the first granite pavements were laid down, which were found so beneficial and in the end economical, that during the next half-century almost all London was thus paved. In 1768 the first Commissioners of Paving, Lighting, and Watching were appointed, and by 1780 Dr. Black states that many streets had been widened, sewers made, that there was a better water supply and less crowding.[1] From this date onward, we are told in the *Encyclopædia Britannica* (art. "London"), a rapid rate of progress commenced, and that since 1785 almost the whole of the houses within the City had been rebuilt, with wider streets and much more light and air. In 1795 the western side of Temple Bar and Snowhill were widened and improved, and soon afterwards Butcher's Row, at the back of St. Clement's Church, was removed. Of course, these are only indications of changes that were going on over the whole city; and, coincident with these improvements, there was a rapid extension of the inhabited area, which, from a sanitary point of view, was of far greater importance. That agglomeration of streets interspersed with spacious squares and gardens, which extends to the north of Oxford Street, was almost wholly built in the period we are discussing. Bloomsbury and Russell Squares and the adjacent streets, occupy the site of Bedford House and grounds, which were sold for building on in 1800. All round London similar extensions were carried out. People went to live in these new suburbs, giving up their city houses to business or offices only. Regent's Park was formed, and Regent Street and Portland Place were built before 1820, and the whole intervening area was soon covered with streets and houses, which for some considerable period enjoyed the pure air of the country. At this time the water supply became greatly improved, and the use of iron mains in place of the old wooden ones, and of lead pipes by which water was carried into all the new houses, was of inestimable value from a sanitary point of view.

Then, just at the same time, began the great improvement in the roads, consequent on the establishment of mail-coaches in 1784. This at once extended the limits of residence for business men, while it facilitated the supply of fresh food to the city.

[1] See Fourth Report of the Royal Commission on Vaccination, Q. 10,917.

In 1801, London, within the Bills of Mortality, was increased in area by almost 50 per cent., with comparatively very little increase of population, owing to the suburban parishes of St. Luke's, Chelsea, Kensington, Marylebone, Paddington, and St. Pancras being then included; and even in 1821 this whole area had only a million inhabitants, and was therefore still thinly peopled, and enjoying semi-rural conditions of life.[1] The slight increase of population from 1801 to 1821 (about 150,000), notwithstanding this extension of area, proves that these suburban parishes were almost wholly peopled from the denser parts of the city, and to a very small extent by fresh immigrants from the country. It is also clear that many city inhabitants must have removed to outlying parishes beyond the Bills of Mortality, in order to explain the very small increase of population in twenty years. This dispersion of the former city population over a much larger suburban area was, in all probability, the most powerful of the various sanitary causes which led to the great diminution of mortality, both general and from the zymotic diseases.[2]

Another very important agency, at about the same time, was the great change in the popular diet that then occurred—the change from bread, beer, and salted meat or fish, to potatoes, tea or coffee, and fresh meat. Dr. Poore tells us that potatoes were first used in hospital diet in 1767.[3] They steadily grew in

[1] These figures are given in the 8th Annual Report of the Registrar-General, and the parishes included are from the *Encyclopædia Britannica*.

[2] I have already repeatedly referred to the vital importance of space, air, and light for healthy living. A few more illustrations may be here given. In his work already quoted, Dr. Poore gives a table of the mortality by measles and whooping-cough of children under five, for the years 1871–80, in the different districts of London, according to density of population. It gives the following results:—

| | Deaths per 100,000 living. |
|---|---|
| Six districts having more than 150 persons per acre . . . | 1157 |
| Seven " " from 100 to 150 " " . . . | 1077 |
| Seven " " " 50 to 100 " " . . . | 968 |
| Eight " " less than 50 " " . . . | 743 |

The general death-rate follows the same law. In Lewisham, Wandsworth, and Hampstead, with densities under 35 per acre, the death-rates are under 15 per thousand; while in Shoreditch, Whitechapel, St. George-in-the-East, and St. Saviour, Southwark, with densities from 185 to 208 per acre, the death-rates are from 20 to 24 per thousand, according to the latest returns of the Registrar-General.

[3] *London from a Sanitary and Medical Point of View.* 1889.

favour, and in the early part of this century had become so common that they almost completely abolished scurvy, the prevalence of which had no doubt rendered other diseases more fatal. At the same time tea became a common beverage. The consumption of tea in England in 1775 was 5,648,000 lbs., and in 1801, 23,730,000 lbs., a more than fourfold increase—a rate which has never been approached in any subsequent twenty-five years. With tea came the more general use of milk and sugar; and it was this, perhaps, that helped to cause the exceptionally rapid decrease of infant mortality. Again, in the same period, the disuse of the city churchyards for interments became general; cemeteries were formed in various parts of the suburbs, till such interments in any part of London were forbidden in 1845, thus removing one more, and not an unimportant source of disease from the more crowded areas.

Now, the various classes of improvements here briefly indicated—those in the city itself, in wider, cleaner, and less obstructed streets, the construction of sewers, and better water supply; the more wholesome food, especially in the use of potatoes and other vegetables, and tea, with its accompanying milk and sugar. becoming common articles of diet; and, most important of all, the spreading out of the population over a much wider area, enabling large numbers of persons to live under far more healthy conditions—all, as we have seen, occurring simultaneously, and effecting this most fundamental change within the half-century from 1775 to 1825, are in their combination amply sufficient to account for that remarkable decrease of mortality, not, as the Royal Commissioners suggest, pre-eminently in small-pox, but in all the more important diseases, which especially characterised this period. This is strikingly shown by Dr. Farr's table printed in the Third Report (p. 198), of which the portion that especially concerns us is here given. It shows us for two periods, 1771–80 and

| | 1771-80 | 1801-10 |
|---|---|---|
| Fourteen infantile diseases . . . . | 1682 | 789 |
| Small-pox . . . . . . . . | 502 | 201 |
| Fevers . . . . . . . . . | 621 | 261 |
| Consumption . . . . . . . | 1121 | 716 |
| Dropsy . . . . . . . . . | 225 | 113 |

1801–10, the deaths per 100,000 living from the more important diseases.

Here we see that, in the thirty years from 1775 to 1805, a change occurred which reduced the mortality from all the chief diseases to half, or less than half, their previous amount. Small-pox no doubt shows the largest decrease; but, as it is a decrease which was mainly effected before vaccination was heard of, that operation can not have been its cause.[1] Now, the remarkable feature of this diminution of mortality is, that *in no similar period between* 1629, *when the Bills of Mortality began, down to the present year, has there been anything like it.* And the same may be said of the causes that led to it. Never before or since has there been such an important change in the food of the people, or such a rapid spreading out of the crowded population over a much larger and previously unoccupied area; and these two changes are, I submit, when taken in conjunction with the sanitary improvements in the city itself, and the much greater facilities of communication between the town and country around, amply sufficient to account for the sudden and unexampled improvement in the general health, as indicated by the great reduction of the death-rate from all the chief groups of diseases, including small-pox.

Now, in the whole of the Final Report, I can find no recognition whatever of the remarkable and exceptional improvement in the general health of London that has been shown to have occurred in the period embracing the end of the last and the beginning of the present centuries; nor of the equally exceptional changes of various kinds, all tending to improved health in the people. And, in view of the facts here adduced, the statement of the Royal Commissioners that "no evidence is forthcoming to show that during the first quarter of the nineteenth century these improvements differentiated that quarter from the preceding quarter or half of the preceding century in any way at all comparable to the extent of the differentiation in respect to small-pox," has, I submit, been shown to be wholly erroneous.

And with respect to the absence of proof of similar changes having occurred in other European countries, which they also urge against the sanitation theory, we hardly need any such

[1] The decrease is probably exaggerated, owing to the confusion of measles with small-pox. Measles shows an increased mortality in the above period from 48 to 94, and as it increased through the whole of the Bills of Mortality, it was probably being slowly differentiated from small-pox.

proof in detail. The very fact of the immediate adoption of vaccination in all the more civilised countries, shows how rapid was the spread of ideas and of customs at that very period. And when we consider, further, that in the last century all the great European cities were at about the same level of filth and unhealthiness with London, and that a century later there is not much difference between them, the probability is in favour of their having all advanced approximately *pari passu*. And with regard to the all-important change in diet and other habits, the same rule applies. The use of potatoes and of tea or coffee, the better water supply, drainage, ventilation, and good roads, were all adopted,—in France and Germany, at all events,—approximately about the same periods as with us. Hence it is not surprising that a similar diminution in general mortality as well as in mortality from zymotic diseases, including small-pox, should have occurred almost simultaneously. The fact, that when we have fairly good statistics, as in Sweden, the great improvement in small-pox mortality is shown to have occurred *before* the introduction of vaccination or before it could have affected more than a small fraction of the population, sufficiently proves that this was the case.

I have now supplied the last piece of confirmatory evidence which the Commissioners declared was not forthcoming; not because I think it at all necessary for the complete condemnation of vaccination, but because it affords another illustration of the curious inability of this Commission to recognise any causes as influencing the diminution of small-pox except that operation. In this, as in all the other cases I have discussed, their Report is founded upon the opinions and beliefs of the medical and official upholders of vaccination; while "the great masses of national experience," embodied in statistics of mortality from various groups of diseases, as well as the well-known facts of the sanitary history of London during the critical half-century 1775–1825, are either neglected, misunderstood, or altogether overlooked.

# CHAPTER XIX

## MILITARISM—THE CURSE OF CIVILIZATION

### 1. Crime and Punishment.

They love the most who are forgiven most;
And when right reason slowly dawns once more
On the wild madness of a moral fiend—
Our brother still and God's beloved child—
There comes a mighty gush of gratitude,
Thawing the hoar-frost of a life of crime,
Breaking the icy barriers of self-love,
While all the loosened rivers of the soul
Spring from their fountains radiant in the light.

—*T. L. Harris.*

The vilest deeds, like poison weeds,
  Bloom well in prison air;
It is only what is good in Man
  That wastes and withers there;
Pale Anguish keeps the heavy gate,
  And the Warder is Despair.

—*The Ballad of Reading Gaol.*

The first half of the century produced much good work that has not been further developed, many bright promises that have not been fulfilled. The great amelioration of the criminal law, by the exertions of Sir Samuel Romilly, Sir James Mackintosh, and other reformers, have not been succeeded by any corresponding reform of our system of punishment as a whole, which still remains thoroughly inhuman and unjust, and opposed to all the admitted principles by which punishment among a civilized people should be regulated. At the beginning of the century about twenty-five offences were punishable with death, including burglary, stealing from a house or shop to the value of 40*s.*, forgery, coining, using old stamps on perfumery and hair powder, sheep and horse-stealing, and many others. Capital punishment for all these minor offences was abolished before the middle of the century; our prisons were greatly improved as regards cleanliness and order; and transportation to Tasmania and the other Australian colonies, with all its cruelties and abuses, had been got rid of. But there we have stopped; and our treatment of criminals, though not outwardly so harsh, is quite as much opposed to the admitted principles which should regulate all punishment as it was before; while its effects are hardly, if at all, less injurious to the criminals, both as regards bodily and mental health, than the old bad system of the last century.

Even Plato and other classical writers laid down the principle that one of the great objects of all punishment is the improvement of the criminal. Beccaria in the last century developed this view of the true rationale of punishment, and all modern students and philanthropists admit it; yet during the whole century we have not made a single step in this direction as regards the treatment of adult prisoners. A cast-iron routine, solitude, and a grinding military despotism under which the best characters often suffer most, now characterises our penal system, which is admitted to have the effect of making the good bad and the bad worse; and further, of rendering it almost impossible for a first offender to escape from a life of crime. There is no classification of offenders; no sympathetic instruction; no attempt to improve the character; no preparation for an honest life; no means afforded the discharged prisoner enabling him to live an honest life. We have, again and again, been shown what modern penal servitude is like, by educated men who have endured it. They all tell us that it is a hell upon earth; that its tendency is to crush out every human feeling or higher aspiration; and that it sends the majority of those who endure it back to the outside world, worse in character and less capable of living honestly than they were before they entered the prison walls. The system is utterly unchristian, utterly opposed to civilization, or philosophy, or common sense; yet it remains in full force in these last years of the century, and neither governments nor legislators seem to think it a matter of sufficient importance to devote the necessary time and study to its radical reform.

It must be admitted, that in our Prison system we see one of the most terrible failures of the boasted civilization of the Nineteenth Century.

In an allied department, the confinement of the insane, there is also much room for reform. Their actual treatment, both in public and private asylums, has undergone enormous improvement during the early part of the century, and is now almost as good as it can be made in large asylums, where there is no possibility of that proper classification, isolation, and individual treatment which are essential to curative success. But the great evil lies in the existence of private asylums, kept for profit by their owners; and in the system by which, on the certificate of two doctors, employed by any relative or friend, persons may be forcibly kidnapped and carried to one of these private asylums, without any public inquiry, and sometimes even without the knowledge or consent of their other nearest relatives, or of those friends who know most about them. The well-known cases of Mrs. Weldon and Mrs. Lowe, prove, that perfectly sane persons may be thus incarcerated, with the possibility of making them insane by association with mad people and all the horrors of a crowded asylum. These two ladies were incarcerated because they were spiritualists; that is, because they held the same beliefs as Sir William Crookes, the Earl of Crawford, Gerald Massey, and myself have held for the last thirty years, and for holding which, to be consistent, we and hundreds of other equally sane persons ought to have been permanently confined as lunatics. The great ability and perfect sanity of those ladies, and their having influential friends, rendered it impossible to keep them permanently confined; but we may be sure that many less able persons have been, and are now, cruelly and unjustly deprived of their liberty, and in some cases are made insane by their terrible surroundings. The great danger of trusting exclusively to professional opinions and statements has been shown in my chapters on hypnotism and vaccination. It is therefore imperative that no person shall be deprived of his liberty, on the allegation by any medical authorities of his insanity. The fact of insanity should be decided, not by the patient's *opinions*, but by his *acts*; and these acts should be proved before a jury, who might also hear medical evidence, before condemnation to an asylum. Asylums for the insane should all belong to public authorities, so that the proprietors and managers should have no pecuniary interest in the continued incarceration of their patients.

So late as 1890 a new and voluminous Lunacy Act was passed, and the public no doubt believe that most of the dangers of the old system are removed. But this is not the case. An examination of this Act shows, that private asylums, kept for profit, remain as before. Doctors' opinions are still all-powerful. Under an "urgency order," on the certificate of *one doctor*, a person may be dragged from his or her home to one of these private asylums, and kept there for seven days, or till a judicial order is obtained, which may sometimes be delayed for three weeks. This judicial order is given by a duly authorized magistrate, on formal application by some person interested, and the certificates of *two* doctors. The magistrate *may* see the alleged lunatic, if he pleases, but he may act on the doctor's and petitioner's statements alone. Whatever inquiry he makes is private; but there is little doubt that in most cases he will act on the medical and other statements before him. Then the alleged lunatic is confined for a year; after that for two years more; then for three years; then for five years, if the medical officer of the asylum reports, before the end of each period, that he is still insane.

And if, either at the first inquiry by the magistrate or afterwards, the patient is declared to be sane, and is discharged, there is no provision for giving the alleged lunatic any information as to the cause of his confinement, or the statements of the medical men, or the person's names who caused him to be confined; so that, really, he is still treated as a possible maniac, and is denied redress if his incarcerators have acted illegally. While confined in one of these private asylums the patient's letters to any official *must* be sent, but letters to any other persons, including his nearest relations or friends, are only sent "at the discretion" of the manager. In like manner the visits of relations or friends require an order from a Commissioner in Lunacy or an official visitor of the asylum; but they are not *obliged* to give such an order, so that if the manager of any private asylum states that it is inadvisable, or that it would be injurious to the patient, the order will probably be refused. It thus appears that an alleged lunatic, once in an asylum, is wholly dependent on the doctors for any chance of getting out again. Everything is in their hands. The patient may be deprived of all communication with friends, either personal or by letter; and though he may see or write to a Commissioner, that will avail him nothing if the medical superintendent either mistakenly believes him to be insane, or has personal reasons for keeping him in the asylum. From beginning to end there is no publicity, no opportunity of disproving any statements that may be made against him, no means of proving his sanity in open court, and subject to the usual safeguards which are accorded to the poorest criminal.

Still more dangerous to liberty is the provision, in Sect. 20 of this Act, that any constable, relieving officer, or overseer, may remove any alleged lunatic to the workhouse, if *he is satisfied* that this is necessary for the public safety or the welfare of the alleged lunatic. It seems hardly credible, but the judges, in a court of appeal, have decided that any of the above named persons *may act on the private information of one person without seeing the alleged lunatic* or giving him any opportunity to state or prove that he is not a lunatic! Yet they did so decide in March 1898. A Mr. Harward quarrelled with his wife, and was rather violent, but did not assault or touch her. Yet she went to the relieving officer and said she was afraid her husband would commit suicide or kill her and the children; and on this statement, without any confirmation and without any personal interview, Mr. Harward was taken by force to the workhouse and confined as a lunatic. Being found perfectly sane, he was soon released; and he then brought an action against the Guardians of Hackney Union for false imprisonment. The jury gave him £25 damages, on the ground that "the relieving officer had not taken reasonable care to satisfy himself that the plaintiff was a dangerous lunatic." But the judges decided on appeal that there was no evidence to show that the officer "acted from any other motive than an honest belief," and therefore he was not liable and the plaintiff had no redress. On such grounds, it is evident, that any passionate or violent person may, on a mere statement of a relative professing to fear injury, without any further inquiry, be captured and confined as a lunatic, and have no redress. This is a mere parody of justice. Every one found to have been confined unjustly, for any cause whatever, should receive an apology and compensation from the authorities concerned, without being left to appeal to the law, at great expense and trouble, and with the chance of the further injustice of a decision against him.

In view of such cases as this, and of the recent scandalous kidnapping of Miss Lanchester; and of the proved danger of founding legislation on the statements and opinions of doctors and officials in the matter of compulsory vaccination, the actual state of our Lunacy laws is a permanent danger to liberty and to the free expression of opinion, and is a disgrace to the closing years of our century.

## 2. The Vampire of War.

Were half the power that fills the world with terror,
Were half the wealth bestowed on Camps and Courts,
Given to redeem the human mind from error,
There were no need for Arsenals and Forts.
The Warrior's name would be a name abhorred!
And every nation that should lift again
Its hand against a brother, on its forehead
Would wear for evermore the curse of Cain!

—*Longfellow.*

Since tyrants by the sale of human life
Heap luxuries to their sensualism, and fame
To their wide-wasting and insatiate pride,
Success has sanctioned to a credulous world
The ruin, the disgrace, the woe of War.

—*Shelley.*

The first half of the nineteenth century was signalised by the abolition of duelling. It had always been illegal, and long been considered to be both absurd and wicked by every advanced thinker; but only when forbidden to military men by the War Office did it entirely disappear among civilians. The same public opinion which caused the disappearance of this form of private war, equally condemns war between nations as a means of settling disputes, often of the most trivial kind; and rarely of sufficient importance to justify the destruction of life and property, the national hatreds, and the widespread misery caused by it. Yet so far from any progress having been made towards its abolition, the latter half of the century has witnessed a revival of the war-spirit throughout Europe; which region has now become a vast camp, occupied by opposing forces greater in numbers than the world has ever seen before. These great armies are continually being equipped with new and more deadly weapons, at a cost which strains the resources even of the most wealthy nations, and by the constant increase of taxation and of debt impoverishes the mass of the people.

The first International Exhibition in 1851, fostered the idea that the rulers of Europe would at length recognise the fact, that peace and commercial intercourse were essential to national well-being. But, far from any such rational ideas being acted on, there began forthwith a series of the most unjustifiable and useless dynastic wars which the world has ever seen. The Crimean War in 1854–55, forced on by private interests, with no rational object in view, and terrible in its loss of life; the Austro-Prussian War in 1866; the French invasion of Mexico, and the terrible Franco-German War, were all dynastic quarrels, having no sufficient cause, and no relation whatever to the well-being of the communities which were engaged in them.

The evils of these wars did not cease with the awful loss of life and destruction of property, which were their immediate results, since they formed the excuse for that inordinate increase of armaments and of the war-spirit under which Europe now groans. This increase, and the cost of weapons and equipments, has been intensified by the application to war purposes of those mechanical inventions and scientific discoveries which, properly used, should bring peace and plenty to all, but which, when seized upon by the spirit of militarism, directly tend to enmity among nations and to the misery of the people.

The first steps in this military development were the adoption of a new rifle for the whole Prussian Army in 1846, the application of steam to our ships of war in 1840, and the use of iron armour for the protection of battleships by the French in 1859. The remainder of the century has witnessed a mad race between all the great powers of Europe, to increase the death-dealing power of their weapons, and to add to the number and efficiency of their armies; while among the maritime powers there has been a still wilder struggle, in which all the resources of modern science have been utilised in order to add to the destructive power of cannon, and both the defensive and the offensive powers of ships. The various new explosives have been utilised in shells, mines, and torpedoes; rifled cannon of enormous size and power have been manufactured; while battleships of 10,000 to 15,000 tons displacement, protected by steel armour from ten inches to twenty inches thick, with enormous engines, often at the rate of a horse-power to every ton, driving the ships at a speed of from twelve to twenty-two knots an hour, have so transformed our fleet that the majority of the ships bear no resemblance whatever to the majestic three-deckers and beautiful frigates with which all our great naval victories were gained, and which formed the bulk of our navy only fifty years ago.

Although the total number of warships and of vessels of all kinds in our fleet, are about the same as they were in the middle of the century, their power for offence and defence, and their cost, are immensely greater. Almost all of them are built of iron or steel, and are full of costly machinery; while the torpedo-boats and torpedo-destroyers are adapted for purposes quite different from those of the smaller vessels of our old fleets. Some of our modern first-class armoured turret-ships cost a million sterling; and yet, as in the case of the *Vanguard* off Kingstown in 1875, and more recently the *Victoria* in the Mediterranean, they may be sent to the bottom by a chance collision with a companion ship. The huge 110-ton guns cost £20,000 each, and the more common 67-ton gun costs £14,000. All the modern guns, as well as their projectiles, are elaborate pieces of machinery, finished with the greatest perfection and beauty; and it makes any thoughtful person sad to see such skill and labour, and so much of the results of modern science, devoted to purposes of pure destruction. The six great powers of Europe now possess about 300 battleships and cruisers, from 2,000 up to near 15,000 tons displacement, and nearly 2,000 smaller vessels, which are able to destroy life and property to an extent probably fifty-fold greater than the fleets of the first half of the century.

But even this vast cost and loss to modern civilization, is surpassed by that of the armies of Europe. The numbers of men have greatly increased; their weapons and equipments are more costly; and the reserve forces to be drawn upon in time of war include almost the whole male adult populations, for whom reserves of arms, ammunition, and all military supplies must be kept ready. Counting only the armies of the six great powers on a peace footing, they amount now to nearly three millions of men; and if we add the men permanently attached to the several fleets, we shall have considerably more than three millions of men in the prime of life withdrawn from productive labour, and devoted, nominally to defence, but really to attack and destruction. This, however, is only a portion of the loss. The expense of keeping these three millions of men in food and clothing, in weapons, ammunition, and all the paraphernalia of war; of keeping in a state of readiness the ships, fortifications, and batteries; of continually renewing the stores of all kinds; of pensions to the retired officers and wounded men, and whatever other expenditure these vast military organisations entail, amounts to an annual sum of more than 180 millions sterling.[1] Now, as the average wages of a working man (or his annual expenditure), considering the low wages and the mode of living in Russia, Italy, Austria, and the other continental states, cannot be more than, say, twelve shillings a week, or thirty pounds a year, an expenditure of 180 millions implies the constant labour of at least six million other men in supporting this monstrous and utterly barbarous system of national armaments. If to this number we add those employed in making good the public or private property destroyed in every war, or in smaller military or naval operations in Europe, we shall have a grand total of about ten millions of men withdrawn from all useful or reproductive work, their lives devoted directly or indirectly to the moloch of war, and who must therefore be supported by the remainder of the working community.

And what a horrible mockery is all this when viewed in the light of either Christianity or advancing civilization! All these nations, armed to the teeth, and watching stealthily for some occasion to use their vast armaments for their own aggrandisement and for the injury of their neighbours, are Christian nations. Their governments, one and all, loudly proclaim their Christianity by word and deed—but the deeds are usually some form of disability or persecution of those among their subjects who are not

[1] This is the amount obtained by adding together the war expenditure of the six great powers, as given in *The Statesman's Year Book* for 1897.

orthodox. Of really Christian deeds there are none—no real charity, no forgiveness of injuries, no help to oppressed nationalities, no effort to secure peace or goodwill among men. And all this in spite of the undoubted growth of the true Christian spirit during the last half-century. This spirit has even ameliorated the inevitable horrors of war; by some regard for non-combatants, by greatly increased care for the wounded even among enemies, and by a recognition of some few rights, even of savage races.

Never, perhaps, have the degrading influences of the war-spirit been more prominent than in the last few years, when all the great Christian powers stood grimly by, while a civilized and Christian people were subjected to the most cruel persecution, rapine and massacre, by the direct orders, or with the consent and approval, of the semi-barbarous Sultan of Turkey. Any two of them had power enough to compel the despot to cease his persecution. Some certainly would have compelled him, but they were afraid of the rest, and so stood still. The excuse was even a worse condemnation than the mere failure to act. Again and again did they cry out, "Isolated action against Turkey would bring on a European war." War between whom? War for what? There is only one answer—"For plunder and conquest." It means that these Christian governments do *not* exist for the good of the governed, still less for the good of humanity or civilization, but for the aggrandisement and greed and lust of power of the ruling classes—kings and kaisers, ministers and generals, nobles and millionaires—the true vampires of our civilization, ever seeking fresh dominions from whose people they may suck the very life-blood. Witness their recent conduct towards Crete and Greece, upholding the most terrible despotism in the world, because each one hopes for a favourable opportunity to obtain some advantage, leading ultimately to the largest share of the spoil. Witness their struggle in Africa and in Asia, where millions of savage or semi-civilized peoples may be enslaved and bled for the benefit of their new rulers. The whole world is now but the gambling table of the six great Powers. Just as gambling deteriorates and demoralizes the individual, so the greed for dominion demoralizes governments. The welfare of the people is little cared for, except so far as to make them submissive tax-payers, enabling the ruling and moneyed classes to extend their sway over new territories and to create well-paid places and exciting work for their sons and relatives. Hence comes the force that ever urges on the increase of armaments and extensions of empire. Great vested interests are at stake; and ever-growing pressure is brought to bear upon the too-willing governments in the name of the greatness or the safety of the Empire, the extension of commerce, or the advance of civilization. Anything to distract attention from the starvation and wretchedness and death-dealing trades at home, and the thinly-veiled slavery in many of our tropical or sub-tropical colonies. The condemnation of our system of rule over tributary states is to be plainly seen in plague and famine running riot in India after more than a century of British rule and nearly forty years of the supreme power of the English government.[1] Neither plague nor famine occur to-day in well-

[1] The Parliamentary Papers recently issued on the Plague in India, reveal an insanitary condition of Calcutta and Bombay (and no doubt of most other Indian cities) which is almost incredible; yet we may be sure that it does not erd on the side

governed communities. That the latter, at all events, is almost chronic in India, a country with an indus-

of exaggeration, because it makes known such an utter disregard for the well-being of the Indian peoples, while taxing them to the verge of starvation, as to be nothing less than criminal. These papers, and the discussion on the Plague in Bombay at the Society of Arts, also illustrate that unreliability of interested official statements which we have seen to be so prominent a feature of the vaccination question.

In January 1897 the Indian Government sent the Director-General of the Indian Medical Service, Dr. Cleghorn, to Bombay, to examine personally into the conditions that led to the outbreak and to recommend the best measures for dealing with it. He made "a thorough investigation of the infected quarters," and this is what he states. About 70 per cent. of the whole native population (about 800,000) live in "chawls" or tenement-houses of various sizes, the largest being six or seven stories high and holding from 500 to 1,000 people each. They consist, on each floor, of a long corridor, with small rooms on each side about 8 feet by 12 feet, each room inhabited by a family, often of 5 or 6 persons. The sanitary arrangements were utterly inadequate, the consequence being that the corridors, especially at the ends, became receptacles of filth of every kind, and were apparently never thoroughly cleaned. But the greatest evil of all was, that these over-crowded tenements were built side by side, often with a space of only three or four feet between them, so that even if the windows were open, in all the lower floors there could be neither adequate light nor ventilation. The privacy of Indian domestic life, however, forbade the opening of these windows, so that practically in half at least of the rooms there was neither light nor ventilation. Added to this, the narrow alleys between the chawls, owing to the inadequacy of other accommodation, were used as refuse pits and open sewers where filth was allowed to accumulate, so that both inside and outside there were masses of disease-breeding matter. Even if the rooms and corridors were kept clean, the darkness, the want of ventilation, and the overcrowding, would be sources of deadly disease. With the superadded filth inside and out and the tropical climate, the absence, rather than the presence, of plague would seem the more extraordinary phenomenon, since the condition of London in the 16th and 17th centuries could hardly

trious people and a fertile soil, is the direct result of governing in the interests of the ruling classes instead

have been so bad. The same Parliamentary Paper (up to March, 1897) contains a Sanitary Inspection Report on Calcutta, which goes much more into detail, and describes a state of things of the most terrible and almost incredible nature. As examples—six men and boys lived and cooked in a room 7 × 7 × 6 feet, which had no window, and with filth and sewage all around. Of another street we read: "The houses are built almost back to back. It would be nearly impossible to squeeze between them; sunlight is so far shut out that, with broad daylight outside the gully, it is absolutely impossible to do more than grope your way within these tenements; rats run about here in the dark as they would at night; a heavy sickening odour pervades the whole place; walls and floors are alike damp with contamination from liquid sewage which lies rotting, and for which there is no escape." There are 8 foolscap pages of this Report, going into even more horrible details; and there can be no doubt that a large portion of it will apply just as well to Bombay as to Calcutta, and thus enable us to realise more fully the condition of the many hundred thousand dwellers in the worse parts of that plague-stricken city. In the discussion that followed the reading of Mr. Birdwood's paper at the Society of Arts, Dr. Simpson, late Health Officer in Calcutta—who had been in Bombay assisting the search parties in the plague-stricken districts—stated that, "bad as the houses were in some parts of Calcutta, he found them infinitely worse at Bombay." On the other hand, Mr. Acworth, late Muncipal Commissioner of Bombay, said that the Bombay "chawls" were not so bad as the Calcutta "bustees," that it was "utterly untrue to say that Bombay was a grossly insanitary town," and that it was really the most sanitary large town in India! But the climax of contradiction is reached by the Rev. A. Bowman, late chaplain of Byculla Gaol, Bombay, who stated in a letter to "The Times," (reprinted in the Journal of the Society of Arts, vol. xlvi., p. 333) that he had known the streets and lanes of Bombay intimately for the last five years, and he says, without fear of contradiction, that such places as were described by the Surgeon-General (Dr. Cleghorn) do not exist! The reverend gentleman referred especially to "chawls" holding 1,000 people, and rooms and corridors which the light of day could not enter; but he appar-

of making the interests of the governed the first and the only object. But in this respect India is no worse off than our own country. The condition of the bulk of our workers, the shortness of their lives, the mortality among their children, and the awful condition of misery and vice under which millions are forced to live in the slums of all *our* great cities, are, in proportion to *our* wealth and *their* nearness to the centre of government, even more disgraceful than the periodic famines of remote India. Both are the results of the same system—the exploitation of the workers for the benefit of the ruling caste—and both alike are among the most terrible failures of the century.

The state of things briefly indicated in this chapter is not progress, but retrogression. It will be held by the historian of the future, to show, that we of the nineteenth century were morally and socially unfit to possess and use the enormous powers for good or evil which the rapid advance of scientific discovery had given us; that our boasted civilization was in many respects a mere surface veneer; and that our methods

ently did not then know that Dr. Cleghorn had made these statements in an official memorandum for the information of the Government of India, or he would hardly have made his contradiction so emphatic.

But what are we to think of a government that has allowed the erection of such tenements in the two chief cities of the empire, and which takes no heed of the most rudimentary principles of sanitation till a visitation of plague compels attention to them? A government which spends millions on railroads, on gigantic armies, on annexations and frontier wars, on colleges and schools, and on magnificent public buildings, while allowing a considerable proportion of the native population to live in such horribly insanitary conditions as to rival the worst plague-infested cities of Europe in the middle ages. And this is modern civilization!

of government were not in accordance with either Christianity or civilization. This view is enforced by the consideration that all the European wars of the century have been due to dynastic squabbles or to obtain national aggrandisement, and were *never* waged in order to free the slave or protect the oppressed without any ulterior selfish ends.

It has been often said that Companies have no souls, and the same is still more true of the Governments of our day.

# CHAPTER XX

## THE DEMON OF GREED

What of men in bondage, toiling, blunted,
  In the roaring factory's lurid gloom?
What of cradled infants, starved and stunted?
  What of woman's nameless martyrdom?
The all-seeing sun shines on unheeding,
  Shines by night the calm, unruffled moon,
Though the human myriads, preying, bleeding,
  Put creation harshly out of tune.

—*Mathilde Blind.*

Are there no wrongs of nations to redress;
No misery-frozen sons of wretchedness;
No orphans, homeless, staining with their feet
The very flag-stones of the wintry street;
No broken-hearted daughters of despair,
Forlornly beautiful, to be your care?
Is there no hunger, ignorance or crime?
O that the prophet-bards of old, sublime,
That grand Isaiah and his kindred just,
Might rouse ye from your slavery to the dust.

—*T. L. Harris.*

One of the most prominent features of our century has been the enormous and continuous growth of wealth, without any corresponding increase in the well-being of *the whole people*; while there is ample evidence to show that the number of the very poor—of those existing with a minimum of the bare necessaries of life—has enormously increased, and many indications that they constitute a larger proportion of the whole population than in the first half of the century, or in any earlier period of our history.

This increase of individual wealth is most clearly shown by the rise and continuous increase of millionaires, who, by various modes, have succeeded in possessing themselves of vast amounts of riches created by others, thus necessarily impoverishing those who did create it. Sixty or seventy years ago a millionaire was a rarity. I well remember, in my boyhood, my father reading in the *Times* an account of the death of a man (a merchant, I think) who had left a fortune of a million, as something altogether marvellous which he had never heard of before. Now, they are to be reckoned by scores, if not by hundreds, in this country, and excite no special remark; while in America, a country having a much larger amount of natural wealth and of human labour to draw upon, they are far more numerous, reaching, it is estimated, about two thousand.

In our own country the annual produce of labour from which the whole expenditure of the people necessarily comes, is estimated at 1,350 millions sterling; and this amount is so unequally divided that *one* million persons among the wealthy receive more than *twice* as much of this income as the *twenty-six* millions constituting the manual labour class. In America the inequality is still greater, there being 4,047 families of the rich who own about five times as much property as 6,599,796 families of the poor.

The causes of this enormous inequality of distribution, and of all the evils that flow from it, are alike in both countries—the practical monopoly of the land and all the mineral wealth it contains, by one section of the wealthy, and of what is usually termed capital by another; resulting in the monopoly by these two classes, who may both be termed capitalists, of all the products of industry and all the industrial applications of science. This arises from the fact that those who have neither land nor capital are obliged to work, at competition wages, for the capitalists; who, for the same reason, have the command of all scientific discovery and all the inventive ability of the nation, and even of the whole civilized world. Hence it has happened that the development of steam navigation, of railroads and telegraphs, of mechanical and chemical science, and the growth of the population, while enormously increasing productive power and the amount of material products—that is, of real wealth—at least ten times faster than the growth of the population, has given that enormous increase almost wholly to one class, comprising the landlords and capitalists, leaving the actual producers of it—the industrial workers and inventors—little, if any, better off than before. If this tenfold increase of real wealth had been so distributed that *all* were equally benefited, then every worker would have had ten times as much of the necessaries and comforts of life, including a greater amount of leisure and enjoyment; while none would have starved, none would have slaved fourteen or sixteen hours a day for a bare existence, none need have had their lives shortened by unwholesome or dangerous occupations; and yet the capitalists and landlords might also have had their proportionate share of the increase. As it is, they have had many times more than their proportionate share; the result being that, if we take the whole of the class of manual labourers, little, if any, of the increase has gone to them.

A number of well-established facts prove this. In the first place, the most recent estimates of Giffen,

Mulhall, and Leoni Levi, gave an average annual income of £77, or almost exactly 30*s.* a week, for each adult male of the working classes. But great numbers of these, including all the skilled mechanics, miners, etc., get considerably more than this, so that the remainder must get less. Now, Mr. Charles Booth puts the "margin of poverty" in London at a guinea a week per family, the test being that less than this sum does not afford sufficient of the absolute necessaries of life—food, clothing, a sanitary dwelling, and ample firing—to keep up health and strength; and he estimates that there are in London about 1,300,000 persons who live below this margin; and if we add to these the inmates of workhouses, prisons, hospitals and asylums, we arrive at the fact that about *one-third* of the total population of London are living miserable, poverty-stricken lives, the bulk of them with grinding, hopeless toil, only modified by the still worse condition of want of employment, with its accompaniments of harassing anxiety and partial starvation. And this is a true picture of what exists in all our great cities, and to a somewhat less degree of intensity over the whole country. There is surely very little indication here of any improvement in the condition of the people. Can it be maintained—has it ever been suggested—that in the early part of the century *more* than one-third of the inhabitants of London did not have sufficient of the bare necessaries of life? In order that there may have been any considerable improvement, an improvement in any degree commensurate with the vast increase of wealth, a full *half* of the entire population of London must then have lived in this condition of want and misery; and I am not aware that any writer has even suggested, much less proved, that such was the case. I believe myself that in *no* earlier period has there been such a large proportion of our population living in absolute want—below "the margin of poverty"—as at the present time; hence there has been *no* improvement in the condition of the mass of miscellaneous unskilled workers, who are now far more numerous than they ever were before. A few reasons for this belief may be given.

Since 1856 the Registrar-General has given the number of deaths in workhouses, hospitals, and other public institutions, for London, and also for England and Wales,[1] and in both areas the proportion of such deaths has been *increasing* for the last thirty-five years. In 1888 the Registrar-General called attention to this portentous increase, which has not yet reached its maximum. The following are the figures, in quinquennial averages, since 1870:—

DEATHS IN PUBLIC INSTITUTIONS IN LONDON.

| Years. | Per cent. of total Deaths. | Years. | Per cent. of total Deaths. |
|---|---|---|---|
| 1861–65 | ... 16·2 | 1881–85 | ... 21·1 |
| 1871–75 | ... 17·4 | 1886–90 | ... 23·4 |
| 1876–80 | ... 18·6 | 1891–95 | ... 26·7 |

In 1861–65, the earliest five years, the proportion was 16·2 per cent. In 1892–96, the latest published, it was 26·9. And what makes this more terrible is, first, that during this period private charity has been increasing enormously; and, secondly, that almost weekly we see proofs of a growing dislike to the workhouse, so that numbers actually die of want rather than apply to the relieving officer. From 1860 to 1885, no less than 130 new charitable organisations

[1] The proportions for England and Wales are about half those for London.

had been established in London, and in the next ten years there were nearly 50 more. Many of these were small and local, but others embraced all London, and have continuously increased in power. Dr. Barnardo's Homes, for example, beginning on a very small scale in 1866, have so increased that 5,000 children who would otherwise be paupers or criminals are supported, educated, and started in life either at home or abroad. And the Church of England Society for Providing Homes for Waifs and Strays, established only in 1882, now supports about 2,000 children. There are in London about forty other institutions of similar character, each supporting from 250 to 1,000 children, and fifty others with a smaller number; besides a large number of almshouses, hospitals, reformatories, homes, and charity schools. And all these institutions are constantly appealing for more funds, because they cannot keep up with the ever-increasing flood of want and misery. Then there is the large amount of relief distributed through the Charity Organisation Society, with the shelters, the farm-colony, and the extensive rescue work of the Salvation Army. And all this work of relief has been going on and ever increasing, while the numbers of those who spend their last years and die in public institutions has also been increasing, not in numbers merely, but in proportion to the total deaths. And in the face of this overwhelming evidence of the *increase* of poverty and misery and starvation, the official apologists for things as they are, most writers for the press, and most politicians, go on declaring that pauperism is decreasing, because, by more strict rules, *out-relief* is reduced or refused altogether; while the better class of the suffering poor prefer starvation or suicide to breaking up their home, however miserable, and enduring the servitude and prison-like monotony of the workhouse.

Suicides have indeed increased most alarmingly, from **1,347** in 1861 to **2,796** in 1895. This is for England and Wales; and the increase in proportion to the population has been from 67 per million to 92 per million. An examination of the records of inquests shows, that either absolute want or the dread of want is a very frequent cause; and as the other evidence just adduced indicates the continuous increase of want, while the ever-increasing struggle in all forms of trade leads to the continual discharge of men and women who from illness or old age are unable to do the same amount of work as the younger and more healthy, the two sets of facts are seen to be connected as cause and effect. If, however, poverty and unmerited want were *decreasing*, and the poor were, decade by decade, becoming *better off*, then the large and continuous increase, for more than thirty years, of deaths by suicide and in public charitable institutions, during the very same time that private charity in varied forms had increased at an altogether unprecedented rate, becomes altogether inexplicable. If poverty had been decreasing, then we should expect the enormously increased and widespread sphere of public charity to have easily overtaken the severer forms of distress; to have reduced the deaths in the workhouse and asylum; to have diminished suicide from the dread of destitution; and to have abolished actual death from starvation in the richest and most charitable city in the world. But the facts are exactly the opposite of all this; and I submit that there is no rational explanation of them other than a continuous increase of the extremest forms of misery and want.

*Illustrations of the Poverty of To-day.*

But these figures, proving the unequal distribution of wealth and the widespread destitution in our midst, however important and expressive to the thinker and the student, do not enable the general reader to realize their full meaning without a few concrete examples of what the poverty of to-day actually is. A few illustrative cases will therefore be given as typical of thousands and hundreds of thousands in every part of our country.

And first, let us hear what the author of the *Bitter Cry of Outcast London* had to say in 1883, the statements in which work, though at first denied or declared to be exaggerated, were proved to be exact by the Commission of Enquiry which followed shortly after. And first as to the places in which the very poor live.

"Few who will read these pages have any conception of what these pestilential human rookeries are, where tens of thousands are crowded together amidst horrors which call to mind what we have heard of the middle passage of the slave ship. To get into them you have to penetrate courts reeking with poisonous gases arising from accumulations of sewage and refuse scattered in all directions and often flowing beneath your feet; courts, many of them which the sun never penetrates, which are never visited by a breath of fresh air, and which rarely know the virtues of a drop of cleansing water. You have to ascend rotten staircases which threaten to give way beneath every step, and which in some places have already broken down. You have to grope your way along dark and filthy passages swarming with vermin. Then, if you are not driven back by the intolerable stench, you may gain admittance to the dens in which these thousands of beings, who belong, as much as you to the race for whom Christ died, herd together." . . .

"Every room in these reeking tenements, houses a family or two. In one room a missionary found a man ill with small-pox, his wife just recovering from her confinement, and the children running about half naked and covered with dirt. Here are seven people living in one underground kitchen, and a little dead child lying in the same room. Here lives a widow and her six children, two of whom are ill with scarlet fever. In another, nine brothers and sisters from twenty-nine years of age downwards, live, eat, and sleep together."

And so the wretched and shameful story goes on, and the author assures us that these are not "selected cases," but that they simply show what is to be found "in house after house, court after court, street after street"; and that the accounts are in no way exaggerated, but are often toned down, because the actual facts are too horrible to be printed.

And next, as to the work by which they live. A woman, trouser-making, can earn one shilling a day if she works *seventeen hours* at it. A woman with a sick husband and a little child to look after, works at shirt-finishing, at 3*d.* a dozen, and can earn barely 6*d.* a day. Another maintains herself and a blind husband by making match boxes at 2¼*d.* a gross, and has to pay a girl 1*d.* a gross to help her. Here is a mother who has pawned her four children's clothes, not for drink, but for coals and food. She obtained only a shilling, and bought seven pounds of coals and a loaf of bread! Think of the agony of distress a mother must have endured before she could do this! And the fifteen years that have passed, notwithstanding the "Royal Commission," leaves it all just as bad as before. This is what Mr. Arthur Sherwell says, in his recently published *Life in West London*, as to the district north of Soho, where there are more than 100,000 persons living below "the margin of poverty."

"Even under normal conditions the pressure of poverty represented by these figures is extreme; but when, as in 1895, the winter is of exceptional severity, the pressure becomes intolerable. Many of the families lived for weeks on soup and bread from the various charitable soup-kitchens in the neighbourhood. Every available article of furniture or clothing was sold or pawned; in some cases the boots were taken off the children's feet and pawned for bread or fuel. A number of families even in the bitterest times of the long frost, lived for days without fire or light, and often with no food but a chance morsel of bread or tea. One family had lived for weeks on bread and tea and dripping. In another room a family was found consisting of the mother and six children (the father had been in the infirmary for seven weeks), who had lived on a pennyworth of bread, a pennyworth of tea, a halfpennyworth of sugar, and a halfpennyworth of milk—*every other day*, and this was got on credit. . . . In a filthy room in another street were found several children entirely naked (this in the severest days of the long frost)! Their mother had been out since morning looking for work. Several cases were found where the family had been without food (sometimes without fire also) for three days." And while all this was going on, and in one street there were 115 adults out of work, 80 of whom had been so from one to nine months, there were, in the same district, between seven and eight thousand paupers in the various workhouse institutions.

As one more example from a different area we have Mrs. Hogg's account of the fur-pullers of South London, in the *Nineteenth Century* of November, 1897:—

"The room is barely eight feet square, and it has to serve for day and night alike. Pushed into one corner is the bed, a dirty pallet tied together with string, upon which is piled a black heap of bedclothes. On one half of the table are the remains of breakfast —a crust of bread, a piece of butter, and a cracked cup, all thickly coated with the all-pervading hairs. The other half is covered with pulled skins waiting to be taken to the shop. The window is tightly closed, because such air as can find its way in from the stifling court below would force the hairs into the noses and eyes and lungs of the workers, and make life more intolerable for them than it is already. To the visitor, indeed, the choking sensation caused by the passage of the hairs into the throat, and the nausea from the smell of the skins, is, at first, almost too overpowering for speech."

Two women work in this horrible place for twelve hours a day, and can then earn only 1*s.* 4*d.*, out of which comes cost of knives and knife-grinding, and fines and deductions of various kinds. In another room one woman kept herself and a daughter of nine by working all day and earning only about 7*s.* 6*d.* a week. When the work was over she was often so exhausted that she threw herself on the bed too tired even to get food. And for these poor people, of whom there are thousands, there is no hope, no future, but a life of such continuous labour, discomfort, and penury, as to be almost unimaginable to ordinary people.

The descriptions now given illustrate the horrible gulf of extreme poverty in which more than a quarter of a million of the people of London constantly live, and into which sooner or later are precipitated almost the whole of the million and a quarter who are permanently living below the poverty line, and to whom illness or want of work brings on absolute destitution. And we must note that none of these writers, who really know the people they write about, impute any considerable proportion of this misery to vice or drink, but to conditions over which the sufferers have no control; while it is certain that both vice and drink are very frequently the consequences of the very conditions of life they are supposed to bring about.

And for this condition of things there is absolutely no suggestion of a remedy by our legislators. Better housing has been *talked* about this twenty years, but if *done*, how would it supply work, or food, or coals, or clothing? The very suggestion that better *houses* is the one thing needed is a cruel mockery, and a confession of impotence and failure.

### *Dangerous and Unhealthy Trades.*

Equally terrible with the amount of want and misery, due mainly to insufficient earnings, want of work, or illness, is the enormous injury to health and shortening of life due to unhealthy and dangerous trades, almost all of which could be made healthy and safe if human life were estimated as of equal value with the acquisition of wealth by individuals.

In Mrs. C. Mallet's tract on *Dangerous Trades for Women*, we find it stated that girls who do the carding in the linen trade lose their health in about twelve years; the very strongest picked men in the alkali works as a rule do not live to be fifty; glass-blowers become prematurely old at forty, and sometimes become blind; in the Potteries deaths from phthisis are three times as numerous as among other workers. But all these trades are inferior in deadliness to the white lead manufactures, in which numbers of girls and women are employed. Some work on for several years without appreciable injury, but the majority suffer greatly in a year or two, many die in a few months, and some in a few weeks or even days. In this trade the percentage of deaths is higher than in any other, and the real amount is never known, because, when the workers become ill they are usually discharged. They then perhaps work for a time at some other employment, perhaps in another place, and if they ultimately die of lead-poisoning or its consequences, their connection with the dangerous trade is lost. The children born of lead-workers usually die of convulsions, and one woman lost eight children in this way. Mr. Robert Sherrard, in his *White Slaves of England*, has given a later and fuller account, perfectly agreeing with Mrs. Mallet's statements published three years earlier; and notwithstanding the abuse and denials by interested parties, all his essential facts are fully borne out by the quotations he now gives in an Appendix, from the reports of several committees, select or departmental, which have inquired into the various trades he has described, together with the evidence from coroners' inquests and other sources. Any one who reads this Appendix alone will be thoroughly convinced of the terrible amount of human suffering and of death resulting from the "dangeous trades" of England, though their total amount can never be fully realized.

And the whole of this destruction of human life and happiness is absolutely needless, since many of the products are not necessaries of life, and all without exception *could* be made entirely harmless if adequate pressure were brought to bear upon the manufacturers. Let every death that is clearly traceable to a dangerous trade be made manslaughter, for which the owners, or, in the case of a company, the directors, are to be punished by imprisonment, *not* as first-class misdemeanants, and ways will soon be found to carry away or utilise the noxious gases, and provide automatic machinery to carry and pack the deadly white lead and bleaching powder; as would certainly be done if the owners' families or persons in their own rank of life were the only available workers.

Even more horrible than the white-lead poisoning is that by phosphorus, in the match-factories. Phosphorus is not necessary to make matches, but it is a trifle cheaper and a little easier to light (and so more dangerous), and is therefore still largely used; and its effect on the workers is terrible, rotting away the jaws with the agonising pain of cancer, often followed by death. Will it be believed in future ages, that this horrible and unnecessary manufacture, the evils of which were thoroughly known, was yet allowed to be carried on to the very end of this century, which claims so many great and beneficent discoveries, and prides itself on the height of civilization it has attained? To what a depth of helplessness must the poor be brought, when young girls eagerly throng to these deadly trades, rather than face the struggle for food and life by other means!

And in the midst of this very pandemonium of want and suffering, the rich are ever becoming more rich, and boast of it. The *City Press* tells us, that the increased profits in the City of London during the ten years from 1880 to 1890 was no less than £30,755,283, and it adds: "This is the best evidence that can be furnished of our commercial prosperity." A million people in London without sufficient food and clothing and fire for a healthy life—but great commercial prosperity! Thousands maimed or racked and tortured to death by dangerous trades—but great commercial prosperity! Those who die paupers' deaths increasing in the ten years from 21 to 26 per cent. of the total deaths—but what of that, when we have great commercial prosperity! The average lives of the lower class of artisans and workers in the unwholesome trades being only 29 years, while that of the upper classes is 55 years—millions thus killed 25 years before their time; but then we have "Great Commercial Prosperity"!

With remarkable foresight, Professor Cairnes, in 1874, wrote, that so long as the workers were dependent on the capitalists for employment "the margin for the possible improvement of their lot is confined within narrow barriers which cannot be passed, and the problem of their elevation is hopeless. As a body they will not rise at all. A few, more energetic or more fortunate than the rest, will from time to time escape, as they do now, from the ranks of their fellows to the higher walks of industrial life, but the great majority will remain substantially where they are. The remuneration of labour, as such, skilled or unskilled, can never rise much above its present level."[1]

[1] *Some Leading Principles of Political Economy*, p. 348.

The result of a quarter of a century more of this dependence, though the capitalists as a class have become enormously richer, is the state of things here imperfectly depicted. And so it must remain till the workers learn what alone will save them, and take the matter into their own hands. The capitalists will consent to nothing but a few small ameliorations, which may improve the condition of select classes of workers, but will leave the great mass just where they are. For without these thousands of struggling, starving humanity, which furnish an inexhaustible reserve of cheap labour, they believe that they cannot go on increasing their wealth; and they systematically oppose all measures which would utilise that labour for the well-being of the labourers themselves, and thus raise wages from the very bottom. This explains why they ignored Mr. Mather's very moderate scheme submitted to the Select Committee on the Unemployed, as well as the far more effectual and practical scheme of Mr. Herbert V. Mills, fully explained in his *Poverty and the State* nine years ago.

A few years before his much-lamented death, that acute yet cautious thinker, the late Professor Huxley, was forced to adopt the conclusions of Professor Cairnes and those here set forth, that our modern system of landlordism and capitalistic competition tends to increase rather than to diminish poverty; and he expressed them in one of those forcible passages which cannot be too often quoted. After declaring that in all great industrial centres there is a large and *increasing* mass of what the French call *la misère*, he goes on:—

"It is a condition in which food, warmth, and clothing, which are necessary for the mere maintenance of the functions of the body in their normal state, cannot be obtained; in which men, women, and children are forced to crowd into dens where decency is abolished, and the most ordinary conditions of healthful existence are impossible of attainment; in which the pleasures within reach are reduced to brutality and drunkenness; in which the pains accumulate at compound interest in the shape of starvation, disease, stunted development, and moral degradation; in which the prospect of even steady and honest industry is a life of unsuccessful battling with hunger, rounded by a pauper's grave. . . . *When the organization of society*, instead of mitigating this tendency, *tends to continue and intensify it*, when a given social order plainly *makes for evil and not for good*, men naturally enough begin to think it high time to try a fresh experiment. I take it to be a mere plain truth that throughout industrial Europe there is not a single large manufacturing city which is free from a large mass of people whose condition is exactly that described, and from a still greater mass who, living just on the edge of the social swamp, are liable to be precipitated into it."[1]

But there are yet other indications of our terribly unhealthy social condition besides poverty, misery, and preventible deaths. The first is the increase of insanity, which is certainly great, though not perhaps so large as the mere increase of the insane population. This increase from 1859 to 1889 was from **1,867** per million in the former year to **2,907** per million in the latter, or more than 50 per cent. faster than the population. But it is alleged that this is mainly due to the accumulation of patients owing to their being

[1] *Nineteenth Century*, February, 1888.

better taken care of than formerly. This, however, is only a supposition, and an improbable one, since it is admitted that in our crowded asylums proper curative treatment is impossible; and the returns of the Registrar-General show that deaths in lunatic asylums are increasing faster than the number of lunatics. (In the seven years 1888 to 1895 the deaths increased 25 per cent.) And in *Chambers' Encyclopædia*, the writer who gives the above explanation also shows immediately afterwards that it only accounts for the smaller portion of the increase. He says, that if we take the newly registered cases each year "we find they have only risen from 4·5 to 6 per 10,000 (or from 450 to 600 per million) in the thirty years." But this is 30 per cent. faster than the population increases; and it may therefore be taken as the admitted amount of the continuous increase of insanity among us.

Closely connected with insanity is suicide, and that this has very largely increased there is no doubt whatever, as the following table, compiled from the Reports of the Registrar-General, will show:—

| Years. | Deaths by Suicide per Million Living. |
| --- | --- |
| 1866–70 | 66·4 |
| 1871–75 | 66·0 |
| 1876–80 | 73·6 |
| 1881–85 | 73·8 |
| 1886–90 | 79·4 |
| 1891–95 | 88·6 |

Dr. S. A. K. Strahan, in his work on *Suicide and Insanity*, states that: "Within certain limits the rate of suicide ebbs and flows with the prosperity of a nation," and he says that it has been *proved* by several continental writers that the death-rate from suicide "rises and falls with the price of bread." The first statement is undoubtedly true, the latter quite untrue. During the whole period included in the above table the price of wheat was falling from 50*s.* 9*d.* in 1859–61 to 32*s.* 10*d.* in 1889–91. The price of bread is of no importance when the conditions of life are such that thousands of people have not the means of buying any food at all. Insanity and suicide are both largely due to want, or the dread of want, as the weekly records of coroners' inquests and the police courts plainly show.

Yet another indication of the deterioration of the people, owing to the unhealthy and unnatural conditions under which millions of them are compelled to live, is afforded by the continuous increase for the last thirty-five years of premature births, and of congenital defects in those which survive. The following table, showing the proportion to 1,000 births, is from the *Fifty-eighth Annual Report* of the Registrar-General, p. xviii. :—

| Years. | Premature Births. | Congenital Defects. |
|---|---|---|
| 1861–65 | 11·19 | 1·76 |
| 1866–70 | 11·50 | 1·84 |
| 1871–75 | 12·60 | 1·85 |
| 1876–80 | 13·38 | 2·39 |
| 1881–85 | 14·18 | 3·23 |
| 1886–90 | 16·15 | 3·39 |
| 1891–95 | 18·42 | 3·87 |

The worst features of this table are the continuous increase it shows, indicating the action of some constant and increasing cause, and the more rapid increase in the latter half of the period, indicating that the conditions are becoming increasingly worse and worse.

It is the common belief that intemperance has

greatly decreased among us, and no doubt that is the case as compared with the early part of the century. But as regards chronic intemperance resulting in death, the Registrar-General's figures show us that for the last thirty years it has been increasing :—

| Years. | | | Deaths from Alcoholism and Delirium Tremens per Million Living. |
|---|---|---|---|
| 1861–65 | | | 41·6 |
| 1866–70 | | | 35·4 |
| 1871–75 | *Males.* | *Females.* | 37·6 |
| 1876–80 | 60·1 | 24 | 42·2 |
| 1881–85 | 66·6 | 31 | 48·0 |
| 1886–90 | 73·6 | 39·2 | 55·4 |
| 1891–95 | 86·6 | 50·2 | 68·2 |

Here the increase began a little later, but it shows the same alarming fact of being much more rapid in the last fifteen years. For the last twenty years the deaths are given for males and females separately, and we find that the death-rates of the latter from this cause have increased with enormous rapidity. While men's deaths from intemperance have increased about 58 per cent. in the twenty years, those of women have increased more than 100 per cent. The causes that lead to this fatal amount of intoxication are various; but no one will deny that the facts here set forth show the existence of something seriously wrong in our social conditions, and that the evil is rapidly increasing.

There is yet one more indication of our deterioration. One of the arguments in favour of national education was, that it would certainly decrease crime. Herbert Spencer told us that it would not have that effect; that there was nothing in educating the intellect to have any effect on the amount of crime, though it might have an effect upon its character. And he seems to have been right. Owing to changes in the classification of offenders, in the nature of their punishment, in the criminal law, and in the practice of the Courts, it is not difficult to obtain figures showing a decrease, as is often done by officials who will not readily admit that our systems of punishment have no reformatory action. But a gentleman who has had a life-long experience of prisons and prisoners, and has made a serious study of the whole subject, arrives at a different conclusion. He tells us that, after a careful examination of all available statistics for the last thirty years, and making all needful corrections for the changes above referred to, he considers it proved that crime has increased, and at a greater rate than the increase of the population for the same period. The result, which he thinks to be as near the truth as can be obtained from prison and criminal statistics, is as follows :—

| Years. | Prison Population. | In Reformatories and Industrial Schools. | Total. |
|---|---|---|---|
| 1860–69 | 127,690 | 6,834 | 134,524 |
| 1870–79 | 154,145 | 17,394 | 171,539 |
| 1880–89 | 170,827 | 25,505 | 196,332 |

Here we have an increase in the average of the first and last ten-year periods amounting to 46 per cent., while the increase of population in the twenty years from 1865 to 1885 is a little less than 30 per cent.[1]

The writer imputes this result to the continued growth of our great cities, which bring together both

[1] The Rev. W. D. Morrison, late H.M. Chaplain at Wandsworth Prison, in the *Nineteenth Century* for June, 1892.

criminals and those who are preyed upon, and by association and opportunity foster the growth of a criminal population. To this cause, however, must be added the increasing severity of the struggle for existence and our cruel and degrading prison system, which together render it almost impossible for first offenders to gain a livelihood by honest labour.

In concluding this brief sketch of the inevitable results of the struggle for existence and for wealth under present social conditions, I call special attention to the fact, that so many converging lines of evidence point in the same direction. The evidence for the enormous increase of the total mass of misery and want is overwhelming, while, that it has increased even faster than the increase of population is, to my own mind, almost equally clear. But when we see that insanity and suicide, deaths from drink, premature births, congenital defects, and the numbers of criminals have all increased simultaneously, we can hardly help seeing a relation of cause and effect, since the accidental coincidence of so many distinct phenomena is highly improbable, and the first of them—the increase of poverty, combined with dangerous or unhealthy occupations—is admitted to be a true cause, if only a contributory one, of all the rest.

But there is yet another inference to be drawn from the facts and figures which have been set forth in this chapter. If we turn to the table of death-rates in public institutions, we find that they not only increase steadily each quinquennium, but that they increase at a more rapid rate in the later than the earlier years. Dividing the period equally, we find that during the first half the death-rate increased by

21, or rather more than a fifth, while in the second half it increased by ·26, or rather more than a fourth. And when we look at the tables showing the amount of suicides, of premature births, of congenital defects, and of deaths from alcoholism, we find that all these also show a much more rapid increase in the latter half, indicating still more clearly the dependence of the latter upon the former.

Now this portentous phenomenon of the *increasing rate* of deterioration of our population is also seen in the rate of increase of individual wealth. Taking the total annual value assessed to Income Tax as the best available indication of individual wealth at different periods, we find the rate of its increase during three periods of fifteen years each to be as follows:—

| Years. | Increase of Income-Tax Assessment. |
|---|---|
| 1850-65 . . | 64·6 per cent. increase in 15 years. |
| 1865-80 . . | 68·6 " " " |
| 1881-96 . . | 82·4 " " " |

This is for the whole of Great Britain and Ireland, and it corresponds with that recent increase of wealth in the City of London which was taken by the writer in the *City Press* to be a gratifying proof of "commercial prosperity."

Here, then, we have direct confirmation of that "increase of want with increase of wealth" which, when propounded as a fundamental fact of modern social systems in Henry George's *Progress and Poverty*, was fiercely denied as utterly unfounded and the very opposite of the truth. The association of the two phenomena is clearly proved by the facts and figures here given; and that association is shown to be not a mere coincidence by the fact that not only the increase, but changes in the rate of increase, are strictly associated; and, yet further, that four separate indications of deterioration which are, partially or wholly, due to poverty, to dread of poverty, or to rapid fluctuations of wealth, also show similar changes in their rate of increase.

We have seen that, in Huxley's opinion, all the terrible social evils which have been briefly summarized in this chapter are due to the existing *organization of society*, and that our present social order "*makes for evil*"; the late Prof. Cairnes was of the same opinion; Frederick Harrison, in 1886, declared that the condition of the actual producers of wealth was then such as to be the condemnation of modern society,[1]—yet it has since then been getting worse, and all our great thinkers—prophets or poets—have condemned it. Carlyle thundered against its iniquities, but with no clear indications of a remedy; Ruskin saw more clearly that a fundamental change in our methods was necessary, and stated clearly, and I believe truly, what the first essential steps of that change must be.[2] Tennyson asks us—

> "Is it well that while we range with Science, glorying in the Time,
> City children soak and blacken soul and sense in city slime?"

John Stuart Mill long since warned us that when great evils are in question small remedies do not produce a small effect, but no effect at all. And Lowell says the same in his exquisite verse,—

[1] See Report of the *Industrial Remuneration Conference*, p. 429.

[2] See *Unto this Last.* Preface.

> "New occasions teach new duties: Time makes ancient good uncouth;
> They must upward still and onward, who would keep abreast of Truth;
> Lo! before us gleam her camp fires! we ourselves must Pilgrims be,
> Launch our *Mayflower*, and steer boldly through the desperate winter sea,
> Nor attempt the Future's portal with the Past's blood-rusted key."

Yet this is exactly what we *have* been doing during the whole century,—applying small plasters to each social ulcer as it became revealed to us—petty palliatives for chronic evils. But ever as one symptom has been got rid of new diseases have appeared, or the old have burst out elsewhere with increased virulence; and it will certainly be considered one of the most terrible and inexplicable failures of the Nineteenth Century that, up to its very close, neither legislators nor politicians of either of the great parties that alternately ruled the nation, would acknowledge that there *could* be anything really wrong while wealth increased as it was increasing. Our ruling classes have suggested nothing, and have done nothing, of any real use. They have made fruitless inquiries into particular phases of the evils that were oppressing the workers, and have continued the application of those small remedies that always have resulted, and always must result, in no permanent benefit to the whole people. They still believe that "the Past's blood-rusted key" will open the portal of future well-being![1]

[1] It is never my practice to condemn evils without suggesting remedies. But in this work it would be out of place to go into them in detail. I give, however, a few suggestions and references in an Appendix to this volume.

# CHAPTER XXI

## THE PLUNDER OF THE EARTH—CONCLUSION

Commerce has set the mark of selfishness,
The signet of its all-enslaving power,
Upon a shining ore, and called it GOLD;
Before whose image bow the vulgar great,
The vainly rich, the miserable proud,
The mob, of peasants, nobles, priests, and kings,
And with blind feelings reverence the power
That grinds them to the dust of misery.

—*Shelley.*

THE struggle for wealth, and its deplorable results, as sketched in the preceding chapter, has been accompanied by a reckless destruction of the stored-up products of nature, which is even more deplorable because more irretrievable. Not only have forest-growths of many hundreds of years been cleared away, often with disastrous consequences, but the whole of the mineral treasures of the earth's surface, the slow products of long-past eons of time and geological change, have been and are still being exhausted, to an extent never before approached, and probably not equalled in amount during the whole preceding period of human history.

In our own country, the value of the coal exported to foreign countries has increased from about three to more than sixteen millions sterling per annum, the quantity being now about thirty millions of tons; and this continuous exhaustion of one of the necessaries of existence is wholly in the interest of landlords and capitalists, while millions of our people have not sufficient for the ordinary needs or comforts of life, and even die in large numbers for want of the vital warmth which it would supply. Another large quantity of coal is consumed in the manufacture of iron for export, which amounts now to about two millions of tons per annum. A rational organization of society would ensure an ample supply of coal to every family in the country before permitting any export whatever; while, if our social organization was both moral and rational, two considerations would prevent any export, the first being that we have duties towards posterity, and have no right to diminish unnecessarily those natural products which cannot be reproduced; and the second, that the operations of coal-mining and iron-working being especially hard and unpleasant to the workers, and at the same time leading to injury to much fertile land and natural beauty, they should be restricted within the narrowest limits consistent with our own well-being.

In America and some other countries, an equally wasteful and needless expenditure of petroleum oils and natural gas is going on, resulting in great accumulations of private wealth, but not sensibly ameliorating the condition of the people at large. Such an excellent light as that afforded by petroleum oil is no doubt a good thing; but it comes in the second grade, as a comfort not a necessity; and it is really out of place till every one can obtain ample food, clothing, warmth, house room, and pure air and water, which are the absolute *necessaries* of life, but which, under the conditions of our modern civilization—more correctly barbarism—millions of people, through no fault of their own, cannot obtain. In these respects we are as the scribes and pharisees, giving tithe of mint and cummin, but neglecting the weightier matters of the law.

Equally disastrous in many respects, has been the wild struggle for gold in California, Australia, South Africa, and elsewhere. The results are hardly less disastrous, though in different ways, than those produced by the Spaniards in Mexico and Peru four centuries ago. Great wealth has been obtained, great populations have grown up and are growing up; but great cities have also grown up with their inevitable poverty, vice, overcrowding, and even starvation, as in the old world. Everywhere, too, this rush for wealth has led to deterioration of land and of natural beauty, by covering up the surface with refuse heaps, by flooding rich lowlands with the barren mud produced by hydraulic mining; and by the great demand for animal food by the mining populations leading to the destruction of natural pastures in California, Australia, and South Africa, and their replacement often by weeds and plants neither beautiful nor good for fodder.

It is also a well-known fact that these accumulations of gold-seekers lead to enormous social evils, opening a field for criminals of every type, and producing an amount of drink-consumption, gambling, and homicide altogether unprecedented. Both the earlier gold-digging by individual miners and the later quartz-mining by great companies, are alike forms of gambling or speculation; and while immense fortunes are made by some, others suffer great losses, so that the gambling spirit is still further encouraged and the production of real wealth by patient industry, to the same extent diminished and rendered less attractive. For it must never be forgotten that the whole enormous amount of human labour expended in the search for and the production of gold; the ships which carry out the thousands of explorers, diggers, and speculators; the tools, implements, and machinery they use; their houses, food, and clothing, as well as the countless gallons of liquor of various qualities which they consume, are all, so far as the well-being of the community is concerned, absolutely wasted. Gold is *not* wealth; it is neither a necessary nor a luxury of life, in the true sense of the word. It serves two purposes only: it is an instrument used for the exchange of commodities, and its use in the arts is mainly as ornament or as an indication of wealth. Nothing is more certain than that the appearance of wealth produced by large gold-production is delusive. The larger the proportion of the population of a country that devotes itself to gold-production, the smaller the numbers left to produce real wealth—food, clothing, houses, fuel, roads, machinery, and all the innumerable conveniences, comforts, and wholesome luxuries of life. Hence, whatever appearances may indicate, gold-production makes a country poor, and by furnishing new means of investment and speculation helps to keep it poor; and it has certainly helped considerably in producing that amount of wretchedness, starvation, and crime which, as we have seen, has gone on increasing to the very end of our century.

But the extraction of the mineral products stored in the earth, in order to increase individual wealth, and to the same extent to the diminution of national well-being, is only a portion of the injury done to posterity by the "plunder of the earth." In tropical countries many valuable products can be cultivated by means of cheap native labour, so as to give a large profit to the European planter. But here also the desire to get rich as quickly as possible has often defeated the planters' hopes. Nutmegs were grown for some years in Singapore and Penang; but by the exposure of the young trees to the sun, instead of growing them under the shade of great forest-trees, as in their natural state, and as they are grown in Banda, they became unhealthy and unprofitable. Then coffee was planted, and was grown very largely in Ceylon and other places; but here again the virgin forests were entirely removed, producing unnatural conditions, and the growth of the young trees was stimulated by manure. Soon there came disease and insect enemies, and coffee had to be given up in favour of tea, which is now grown over large areas both in Ceylon and India. But the clearing of the forests on steep hill slopes to make coffee plantations produced permanent injury to the country of a very serious kind. The rich soil, the product of thousands of years of slow decomposition of the rock, fertilized by the humus formed from decaying forest trees, being no longer protected by the covering of dense vegetation, was quickly washed away by the tropical rains, leaving great areas of bare rock or furrowed clay, absolutely sterile, and which will probably not regain its former fertility for hundreds, perhaps thousands, of years. The devastation caused by the great despots of the middle ages and of antiquity, for purposes of conquest or of punishment, has thus been reproduced in our times by the rush to obtain wealth.

Even the lust of conquest, in order to secure slaves and tribute and great estates, by means of which the ruling classes could live in boundless luxury, so characteristic of the early civilizations, is reproduced in our own time. The great powers of Europe are in the midst of a struggle, in order to divide up the whole continent of Africa among themselves, and thus obtain an outlet for the more energetic portions of their populations and an extension of their trade. The result, so far, has been the sale of vast quantities of rum and gunpowder; much bloodshed, owing to the objection of the natives to the seizure of their lands and their cattle; great demoralization both of black and white; and the condemnation of the conquered tribes to a modified form of slavery. Comparing our conduct with that of the Spanish conquerors of the West Indies, Mexico, and Peru, and making some allowance for differences of race and of public opinion, there is not much to choose between them. Wealth, and territory, and native labour, were the real objects of the conquest in both cases; and if the Spaniards were more cruel by nature, and more reckless in their methods, the results were much the same. In both cases the country was conquered, and thereafter occupied and governed by the conquerors frankly for their own ends, and with little regard to the feelings or the material well-being of the conquered. If the Spaniards exterminated the natives of the West Indies, we have done the same thing in Tasmania, and almost the same in temperate Australia. And in the estimation of the historian of the future, the Spaniards will be credited with two points in which they surpassed us. Their belief that they were really serving God in converting the heathen, even at the point of the sword,was a genuine belief shared by priests and conquerors alike—not a mere sham, as is ours when we defend our conduct by the plea of introducing the "blessings of civilization." And, in wild romance, boldness of conception, reckless daring, and the successful achievement of the well-nigh impossible, we are nowhere when compared with Cortez and his five hundred Spaniards, who, with no base of supplies, no rapid steam communication, no supports, imperfect weapons and the ammunition they carried with them, conquered great, populous, and civilized empires. It is quite possible that both the conquest of Mexico and Peru by the Spaniards, and our conquest of South Africa, may have been real steps in advance, essential to human progress, and helping on the future reign of true civilization and the well-being of the human race. But if so, we have been, and are, unconscious agents, in hastening that great—

"far-off, divine event
To which the whole creation moves."

We deserve no credit for it. Our aims have been, for the most part, sordid and selfish; and if, in the end, all should work out for good, as no doubt it will, much of our conduct in the matter will yet deserve, and will certainly receive, the severest condemnation.

Our whole dealings with subject races has been a strange mixture of good and evil, of success and failure, due, I believe, to the fact that, along with a genuine desire to do good and to govern well, our rule has always been largely influenced, and often entirely directed, by the necessity of finding well-paid places for the less wealthy members of our aristocracy, and also by the constant craving for fresh markets by the influential class of merchants and manufacturers. Hence the enormous fiscal burdens under which the natives of our Indian Empire continue to groan; hence the opium monopoly and the salt tax; hence the continued refusal to carry out the promises made or implied on the establishment of the Empire, to give the natives a continually increasing share in their own government, and to govern India solely in the interest of the Indians themselves.

It is the influence of the two classes above referred to that has urged our governments to perpetual frontier wars, and continual extensions of the Empire, all adding to the burdens of the Indian people. But our greatest mistakes of all are, the collection of revenue in money, at fixed times, from the very poorest cultivators of the soil; and the strict enforcement of our laws relating to landed property, to loans, mortgages, and foreclosures, which are utterly unsuited to the people, and have led to the most cruel oppression, and the transfer of numbers of small farms from the ryots to the money-lenders. Hence, the peasants become poorer and poorer; thousands have been made tenants instead of owners of their farms; and an immense number are in the clutches of the money-lenders, and always in the most extreme poverty. It is from these various causes that the periodical famines are so dreadful a scourge, and such a disgrace to our rule.[1] The people of India are industrious, patient, and

[1] These facts, together with our most cruel and wicked robbery of the rayats, or cultivators, constituting three-fourths of the entire population, by changing the *land-tax* to a *rack-rent* as exorbitant and impossible of payment as these of the worst Irish and Highland landlords, have been long known, and have been again and again urged by the most experienced Indian

frugal in the highest degree; and the soil and climate are such that the one thing wanted to ensure good crops and abundance of food is water-storage for irrigation, and absolute permanence of tenure for the cultivator. That we have built costly railways for the benefit of merchants and capitalists, and have spent upon these and upon frontier-wars the money which would have secured water for irrigation wherever wanted, and thus prevented the continued recurrence of famine whenever the rains are deficient, is an evil attendant on our rule which outweighs many of its benefits.

The final and absolute test of good government is the well-being and contentment of the people—*not* the extent of empire or the abundance of the revenue and the trade. Tried by this test, how seldom have we succeeded in ruling subject-peoples! Rebellion, recurrent famines, and plague in India; discontent, chronic want and misery; famines more or less severe, and continuous depopulation in our sister-island at home—these must surely be reckoned among the most terrible and most disastrous failures of the Nineteenth Century.

> "Hear then, ye Senates! hear this truth sublime,
> They who allow Oppression share the crime."

administrators as the fundamental cause of all Indian (as they are of all Irish) famines. But, quite recently, they have been again described, with admirable lucidity and almost unnecessary moderation, by Sir William Wedderburn, whose great experience in India as a District Judge, and long study of the subject, constitute him one of the first authorities. See a series of articles in the periodical *India* for February, March, May, and June 1897. A reprint of the whole under the very appropriate title—"The Skeleton at the (Jubilee) Feast," has been sent to all members of the House of Commons; and they should be read by every one who wishes to comprehend the terrible misgovernment of our Indian Empire.

## Concluding Remarks

We are now in a position to form some general estimate of progress and retrogression during the Nineteenth Century, and to realize to some extent what will be the verdict of the future upon it. We have seen that it has been characterised by a marvellous and altogether unprecedented progress in knowledge of the universe and of its complex forces; and also in the application of that knowledge to an infinite variety of purposes, calculated, if properly utilized, to supply all the wants of every human being, and to add greatly to the comforts, the enjoyments, and the refinements of life. The bounds of human knowledge have been so far extended that new vistas have opened to us in directions where it had been thought that we could never penetrate, and the more we learn the more we seem capable of learning in the ever-widening expanse of the universe. It may be truly said of men of science, that they have now become as gods knowing good and evil; since they have been able not only to utilize the most recondite powers of nature in their service, but have in many cases been able to discover the sources of much of the evil that afflicts humanity, to abolish pain, to lengthen life, and to add immensely to the intellectual as well as to the physical enjoyments of our race.

But the more we realize the vast possibilities of human welfare which science has given us, the more we must recognise our total failure to make any adequate rational use of them. With ample power to supply to the fullest extent necessaries, comforts, and even luxuries for all, and at the same time allow ample leisure for intellectual pleasures and æsthetic enjoyments, we have yet so sinfully mismanaged our social economy as to give unprecedented and injurious luxury to the few, while millions are compelled to suffer a lifelong deficiency of the barest necessaries for a healthy existence. Instead of devoting the highest powers of our greatest men to remedy these evils, we see the governments of the most advanced nations arming their people to the teeth, and expending much of their wealth and all the resources of their science, in preparation for the destruction of life, of property, and of happiness.

With ample knowledge of the sources of health, we allow, and even compel, the bulk of our population to live and work under conditions which greatly shorten life; while every year we see from 50,000 to 100,000 infants done to death by our criminal neglect.

In our mad race for wealth, we have made gold more sacred than human life; we have made life so hard for the many, that suicide and insanity and crime are alike increasing. With all our labour-saving machinery and all our command over the forces of nature, the struggle for existence has become more fierce than ever before; and year by year an ever-increasing proportion of our people sink into paupers' graves.

Even more degrading, and more terrible in its consequences, is the unblushing selfishness of the greatest civilized nations. While boasting of their military power, and loudly proclaiming their Christianity, not one of them has raised a finger to save a Christian people, the remnant of an ancient civilization, from the most barbarous persecution, torture, and wholesale massacre. A hundred thousand Armenians murdered or starved to death while the representatives of the great powers coldly looked on—and prided themselves on their unanimity in all making the same useless protests—will surely be referred to by the historian of the future, as the most detestable combination of hypocrisy and inhumanity that the world has yet produced, and as the crowning proof of the utter rottenness of the boasted civilization of the Nineteenth Century.

When the brightness of future ages shall have dimmed the glamour of our material progress, the judgment of history will surely be, that the ethical standard of our rulers was a deplorably low one, and that we were unworthy to possess the great and beneficent powers that science had placed in our hands.

But although this century has given us so many examples of failure, it has also given us hope for the future. True humanity, the determination that the crying social evils of our time shall *not* continue; the certainty that they *can* be abolished; an unwavering faith in human nature, have never been so strong, so vigorous, so rapidly growing as they are to-day. The movement towards socialism during the last ten years, in all the chief countries of Europe as well as in America, is the proof of this. This movement pervades the rising generation, as much in the higher and best educated section of the middle class as in the ranks of the workers. The people are being educated to understand the real causes of the social evils that now injure all classes alike, and render many of the advances of science curses instead of blessings. An equal rate of such educational progress for another quarter of a century, will give them at once the power and the knowledge required to initiate the needed reforms.

The flowing tide is with us. We have great poets, great writers, great thinkers, to cheer and guide us; and an ever-increasing band of earnest workers to spread the light and help on the good time coming. And as this Century has witnessed a material and intellectual advance wholly unprecedented in the history of human progress, so the Coming Century will reap the full fruition of that advance, in a moral and social upheaval of an equally new and unprecedented kind, and equally great in amount. That advance is prefigured in the stirring lines of Sir Lewis Morris, with which I may fitly close my work:—

"There shall come, from out this noise of strife and groaning,
A broader and a juster brotherhood,
A deep equality of aim, postponing
All selfish seeking to the general good.
There shall come a time when each shall to another,
Be as Christ would have him, brother unto brother.

"There shall come a time when brotherhood grows stronger
Than the narrow bounds which now distract the world;
When the cannon's roar and trumpets blare no longer,
And the ironclad rusts and battle-flags are furled;
When the bars of creed and speech and race, which sever,
Shall be fused in one humanity for ever."

# APPENDIX

## The Remedy for Want in the Midst of Wealth

> The end of Government is to unfold
> The Social into harmony, and give
> Complete expression to the labouring thought
> Of universal genius; first to feed
> The body, then the mind, and then the heart.
> —*T. L. Harris.*

> New Times demand new measures and new men;
> The world advances, and in time outgrows
> The laws that in our fathers' days were best.
> —*Lowell.*

The experience of the whole century, and more especially of the latter half of it, has fully established the fact that, under our present competitive system of capitalistic production and distribution, the continuous increase of wealth in the possession of the capitalist and landowning classes is *not* accompanied by any corresponding diminution in the severity of misery and want or in the numbers of those who suffer from extreme poverty, rendered more unendurable by the presence of the most lavish waste and luxury on every side of them. Even the most cautious writers who really look at the *facts*, are compelled to admit so much as this; but, as I have shown, the actual facts prove more than this. They show clearly that with the increase of wealth there has been a positive and very large *increase* of want; while if we take account of *all* the facts, and without prejudice or prepossession consider what they really imply, we are driven to the conclusion that, during the latter half of this most marvellous of all the centuries, while science has been enlarging man's power over nature in a hundred varied ways, resulting in possibilities of wealth-production a hundred-fold that of any preceding century, the direst want of the bare necessaries of life has seized upon, not only a greater absolute number, but a larger proportion of our population; and this has happened notwithstanding an increase of charity and benevolent work among the poor which is equally unexampled.

Many of our greatest writers and clearest thinkers have observed these facts, and have plainly declared that our social system has broken down. The number of those who see this is increasing daily; and the public conscience is being aroused by the heart-rending misery and suffering of millions of those who work, or beg to be allowed to work, in order to produce comforts and luxuries for others while living in poverty, hunger, and dirt themselves. I take it for granted, that we shall not much longer permit this social hell to surround us on every side without making some strenuous efforts to abolish it. To do this with the slightest chance of success we must recognise the absolute inefficiency of the old methods of charity and other small ameliorations, except as admittedly temporary measures; and we must devote ourselves to work on new lines which must be fundamental in their nature and calculated to remove the *causes* of poverty.

I have myself indicated those lines in an address to the Land Nationalisation Society in 1895, reprinted with alterations and addition in *Forecasts of the Coming Century*, of which it forms the first article, under the title *Reoccupation of the Land*. The principle is, briefly, the Organisation of Labour, in Production, *for the Consumption of the Labourers*. Nobody has attempted, seriously, to show why this should not be done. Even if the land and stock necessary to start each such co-operative colony were given free, it would be the wisest and most profitable public expenditure ever made, because it is certain to abolish all unmerited poverty by absorbing *all the unemployed*. I have shown by sufficient examples the enormous economies of such organization of labour—economies so great and acting in so many directions that it is quite certain to result, not only in a subsistence for the workers, but in an abundance of all the necessaries, comforts, and rational enjoyments of life.

Just consider for a moment. The workers of the country, very imperfectly organised by the capitalists in their own interests, *do* actually produce every year all the wealth that is consumed, including not only necessaries and comforts, but an enormous quantity of luxuries consumed only by the wealthy.

All these workers when in full work *do* earn enough to live on, and many of them to live comfortably, although they are paid less than half, often only a quarter of the value of their work in the finished article. It is only because the value they add to the product is many times more than the wages they receive, that there is a surplus sufficient to give a profit to the capitalist-manufacturer and to two or three middle-men, to pay for railway-carriage, for travellers, and for advertisements, as well as for loss upon unsaleable goods. *All* these expenses would be saved when almost everything was made to be consumed on the spot by the producers themselves, only a few surplus products being sold in the nearest market to pay for some foreign luxuries. How could such an organisation fail to succeed? If it is said that the unemployed are not first-rate workmen, we reply, second- or even third-rate men will do very well. Average mechanics—carpenters, masons, plumbers, tanners, tailors, shoemakers, spinners, weavers, agricultural labourers, etc.—will be able to build second-class houses, make second-class clothing, and produce plain food. Again, why not? If every kind of trade and manufacture can be carried on and well managed by public companies, whose shareholders know nothing of the business, why not by the local authorities? Every company has to compete with other companies, and with great capitalists in the sale of its products. Here there would be no competition, as the great bulk of the products would be consumed by the producers themselves, and in some cases exchanged for the products of other similar settlements when it is found to be beneficial to do so. Why, then, is this not done? Why is it nowhere attempted? There is really only one answer. Manufacturers and capitalists *are afraid it would succeed.* They *know,* in fact, that it would only succeed too well, that it would render those who are now unemployed self-supporting; and, by abolishing the spur of starvation, or the dread of starvation, would raise wages all round. Hence, so long as we have capitalist governments, and the workers are so blind as to send manufacturers and capitalists and lawyers to misrepresent them in parliament, a really effective remedy will not be tried.

But will advanced thinkers and the educated workers continue much longer to permit myriads to suffer penury that a few may get rich? for that is really what it comes to. The mere consideration that the powers of production are now practically unlimited, and that not only enough for every human being, but far more than could possibly be consumed, can be produced by the machinery and labour now in existence, shows how cruel and unnecessary is the system that condemns so many men, and women, and children, either to long hours of grinding labour, or to idleness and its attendant want and misery.

The ingenious sophistries of modern writers, from the point of view of the competitive and capitalistic system as an absolute fundamental fact, have rendered it difficult for most people to comprehend the reason of the paradox, that with an enormous increase of wealth and of power of producing all commodities there should be a corresponding perpetuation, or even increase of poverty. We owe it to an American writer to have cleared up this difficulty more completely and more intelligibly than has ever been done before; and I strongly recommend those who wish to understand how it is that our capitalistic individualism *necessarily* produces and perpetuates poverty, to read chapter xxii. of Mr. Edward Bellamy's new book *Equality*, entitled "Economic Suicide of the Profit System." Although the form of this chapter is not perhaps the best, being that of a school-examination, it is, nevertheless, an admirably reasoned discussion of the problem, and is, in my judgment, absolutely conclusive. Chapter xxvi. extends the discussion to the effects of foreign trade, both free and protectionist; and shows that under our capitalist and competitive system this only further intensifies the evil as regards the poverty of the masses. Another chapter (xxiii.), entitled "The Parable of the Water Tank," is an amusing illustration of the absurdity of our system, in which a superabundance of all the necessaries of life produced by the labour of the people, actually increases the want and starvation of the same people!

Seeing, then, that the *actual facts* of the case, at the end of our Century of ever-increasing capacity of wealth-production, are in complete accordance with its *necessary results* logically reasoned out from the premises of competitive capitalism, we are bound as rational beings to get rid of this system with as little disturbance as possible, and, therefore, by some process of evolution; but, nevertheless, in such a way as at once to remedy its most cruel and disgraceful effects. The method I have suggested is one of the least revolutionary, while it is both the easiest and the most effective; and during its gradual extension experience will be gained as to the best methods of carrying it out over the whole country.

### *How to Stop Starvation.*

But, till some such method is demanded by public opinion, and forced upon our legislators, the horrible scandal and crime of men, women, and little children, by thousands and millions, living in the most wretched want, dying of actual starvation, or driven to suicide by the dread of it—MUST BE STOPPED! I will therefore conclude with suggestions for stopping this horror *at once*; and also for obtaining the necessary funds, both for this temporary purpose and to carry out the system of co-operative colonies already referred to.

The only certain way to abolish starvation, not when it is too late but in its very earliest stages, is *free bread*. I imagine the outcry against this—"pauperisation! fraud! loafing!" etc., etc. Perhaps so; perhaps not. But if it *must* be so, better a hundred loafers than a thousand starving; and if my main proposal or something equally effective is adopted, the loafers will soon be disposed of. I have thought over this plan of free bread for a couple of years, and I now believe that all the difficulties may be easily overcome. In the first place, *all* who want it, all who have not money to buy wholesome food, must be enabled to get this bread with the minimum of trouble. There must be no *tests*, like those for poor-law relief. A decent home with good furniture and good clothes must be no bar; neither must the possession of money, if that money is required for rent, for coals, or for any other necessaries of life. The bread must be given to *prevent* injurious penury, not merely to alleviate it. Whenever a man (or woman) is out of work from no fault of his own, however good wages he earns when in work, he must have a claim to bread. The bread is *not* to be charity, *not* poor-relief; but a rightful claim upon society for its neglect to so organize itself that *all* without exception who have worked, and are willing to work, or are unable to work, may at the very least have food to support life.

Now for the mode of obtaining this bread. All local authorities shall be required to prepare bread-tickets duly stamped and numbered, of a convenient form, with coupons to be detached, each representing a 4-lb. loaf. These tickets are to be issued in suitable quantities to every policeman, to all the clergy of every denomination, to all medical men, and to such other persons as may be willing to undertake their distribution and are considered to be trustworthy. Any person in want of food, on applying to any of these distributors, is to be given a coupon for one loaf (initialled or signed by the giver) *without any question whatever.* If the person wants more than one loaf, or wishes to have one or more loaves a day for a week or a month, he or she must give name and address. The distributor, or some deputy, will then pay a visit during the day, ascertain the facts, give a suitable number of bread-tickets, and, if needful, as in case of sickness or delicate health, obtain further relief from charitable persons or from any funds available for the purpose.

Now, there are only two possible objections to this method of temporarily stopping starvation while more permanent measures are preparing. The first is, that it would pauperise; the next, that, as wages tend to sink to the minimum for bare subsistence, it would still further lower wages, so that it would then become needful to give coals free, and a little later rent free, till wages were reduced to the Scriptural penny a day, and the whole of the unskilled workers had to be supported. The first objection is absurd; because the effect of this free bread would be to check and almost abolish pauperisation. It would enable the home to be kept up; it would prevent that cruel mockery of the present poor-law system that the home must be denuded or given up, the children's clothes pawned, all self-respect lost, *before* relief is given. The second objection, if valid, would be the strongest condemnation of our actual competitive wage-system. But it is not valid. It is the pressure of absolute hunger, of the still more cruel pang of seeing their children pining for want of bread, that makes men and women consent to work for anything they can get, and gives all the power to the sweater's trade. The being-able to hold out a week or a month would give strength to the poor half-starved women and children now working their lives out in misery and destitution. It would give them power and time to bargain. In each shop or factory they could combine. They could *afford to strike* against oppression, which they dare not do now, and the result would be a rise, not a fall of wages. But, for some persons, that will be an equal objection; and as no one can tell *exactly* what would happen except that *starvation would be abolished,* perhaps it is simpler to ignore all such theoretical and imaginary evils. Let us first stop the starvation, and leave other difficulties to be dealt with as they arise.

*How to get the Funds.*—This question ought not to require asking, in a country where there is such enormous accumulated wealth in the hands of individuals that a large part of it is absolutely useless to them, gives them no rational pleasure, and is, really and fundamentally, the *cause* of the very poverty we seek to abolish.

There are now in Great Britain sixty-six persons whose incomes from "trades and professions" are £50,000 a year and upwards. The total amount of the sixty-six incomes is £5,632,577, so that the *surplus*, over £50,000 a year each, amounts to £2,332,577 a year. Up to the end of the last century it is probable that no one person in Great Britain had an income of £50,000 a year. It would then have been considered what Dr. Johnson termed "wealth beyond the dreams of avarice," and even to-day it is far beyond what is sufficient for every luxury which one family ought to have or ought to want. Surely, for the one purpose of giving BREAD to those who need it, to save MILLIONS from insufficiency of food culminating in absolute starvation, there can be few of these sixty-six who, when appealed to by the humanity, by the intellect, and by the religion of the nation, will refuse to give up this enormous superfluity of wealth to the bread fund, to be taken charge of, perhaps, by the Local Government Board, and administered, on the principles here suggested, by the local authorities. For those who refuse, there will be the scorn and contempt of all good men. In the burning words of Scott,—

> "High though his titles, proud his name,
> Boundless his wealth as wish can claim,
> Despite those titles, power, and pelf,
> The wretch, concentred all in self,
> Living, shall forfeit fair renown,
> And doubly dying, shall go down
> To the vile dust from whence he sprung,
> Unwept, unhonoured, and unsung."

But the above-named amount is only a part, a very small part, of the wealth that is immediately available. There are sitting in the House of Lords sixty peers who hold possession of land producing a rental of over £50,000 a year each. The sum total of these sixty rentals is £5,405,900, so that the amount of the surplus is £2,405,000 a year, and as the average rental is something over a pound an acre, this surplus represents considerably more than two million acres of land. The owners of this surplus land should also be invited to make it over to the nation, to be used, temporarily, for the bread fund, but ultimately for the establishment of the co-operative colonies. Surely these sixty noble lords will not refuse, from their great superfluity, to return a portion to the nation, for the use of those workers who give to the land all its rental value!

But these two surplus revenues, amounting to more than four and a half millions a year, are over and above the enormous revenues derived from the great London estates. Some of these would be wholly available as surplus, since their owners possess incomes of £50,000 a year from other sources; while in other cases the total incomes would be brought to a higher amount than £50,000 by the addition of the London property. There is thus available a fund of at least six or seven millions a year, without reducing any rich man's income below £50,000 a year.

But we should not wish to shut out from this great act of restitution to the nation those persons who possess the comparatively moderate wealth of from £10,000 a year upwards, who might be invited to contribute 10 per cent., 20 per cent., 30 per cent., or 40 per cent. of their surplus over the same number of thousands in their income; and this would certainly produce another million or two million per annum, as there are over a thousand persons in this class with an average income of about £18,000 per annum.

It is estimated that two pounds of bread a day is a full average for the consumption per head, even if no other food is available. The cost of this at 5*d.* the four-pound loaf would be £3 16*s.* a year, so that to supply a million people the *whole year* would require £3,800,000. This might be enough, or there might be a demand for double this; but the very fact of there being so large a want of mere bread would incite to the adoption of permanent means by which *all* could be rendered at least self-supporting; and for this purpose the two millions of acres of land would be at once available as a beginning.

It will probably be objected that *none* of these millionaires will give up their surplus wealth, however piteously we may appeal to them in the name of the suffering millions, from whose labour every pound has been derived, and without whose labour they themselves would be reduced to destitution. Perhaps it may be so. But, if so it be, the people will know the characters of those whom they have to deal with, and will be driven to use their power as voters to obtain by the forms of law what they have not been able to obtain by appeals to either the mercy or the justice of these rich men—who, while calling themselves Christians, will not part with their superfluity of gold and land even to give bread to the poor and needy, and to save widows and the fatherless from misery and starvation.

The means to do this is plain. They must vote for no candidate who will not promise to support—first, a progressive income-tax on that portion of all incomes above £10,000, rising to 100 per cent. on the surplus above £50,000, as here suggested; and, secondly, to support a corresponding or even larger increase in the death duties. The law now permits a man to disinherit his children, or other legal heirs, whenever he chooses; and in thus permitting him recognises the important principle that *no one has an indefeasible claim to succeed to any property whatever!* For great public purposes, therefore, the State may justly declare itself the heir to any proportion of the property, or even to the whole property of deceased persons. But the State would at the same time recognise the duty—which the owner of property does not always recognise—of providing for all persons dependent on the deceased, either by means of an ample annuity for those past middle life, or by a suitable education and start in life for younger relatives or dependents, and for children.

In this way ample funds would be available for the various purposes here suggested, without really injuring any one. These purposes—the abolition of starvation, penury, and the degraded life of millions—are the greatest and most important which any government can undertake, and should, *now*, constitute its first duty. They are the essential first step to any really effective social advance; and if all earnest reformers of every class would unite their forces, their efforts would soon be crowned with success. I have done what I can to prove the utter break-down of our present state of social disorganization —a state which causes all the advances in science and in our command over the forces of nature to be absolutely powerless to check the growth of poverty in our midst. Every attempt to salve or to hide our social ulcers has failed, and must continue to fail, because those ulcers are the necessary product of Competitive Individualism.

I therefore call upon all earnest and thinking men and women to devote their energies to advocating those more fundamental changes which both theory and experience prove to be needed, and which alone have any chance of success.

For now—thought oft mistaken, oft despairing,
  At last, methinks, at last I see the dawn;
At last, though yet a-faint, the awakening nations
  Proclaim the passing of the night forlorn;
Soon shall the long-conceived child of Time
Be born of Progress—soon the morn sublime
Shall burst effulgent through the clouds of Earth,
And light Time's greatest page—O Right, thy glorious birth!

—*J. H. Dell.*

# INDEX

R

Butler & Tanner, The Selwood Printing Works, Frome, and London.

## DIAGRAM 1.

### London Death-rates per Million Living from 1760 to 1896.

The Upper line shows rates of Death from All Causes.

The Middle line shows rates of Death from Zymotic Diseases, including Measles, Fevers, Whooping-cough, and Diphtheria.

The Lower line (shaded for distinctness), Small-pox.

The blank four years, 1834-8, are omitted because they are the last of the old "Bills of Mortality," and are considered to be very imperfect.

From 1838 onwards is the period of complete Registration.

Each ten years is indicated at the bottom and top of the diagram.

The figures at the sides and centre show the mortality per million.

The Upper line (total mortality) is on a smaller vertical scale, and is brought lower down to allow of its being included in the diagram.

*Authorities.*

The lines in the diagram from 1760 to 1834 are calculated from the figures given in the Second Report, pp. 289-91, with those for other diseases from Dr. Creighton's *History of Epidemics in Britain*; the population at the different periods being taken from the best available sources (Maitland, and the 8th Report of the Registrar-General). The later portion is entirely from the Reports of the Registrar-General.

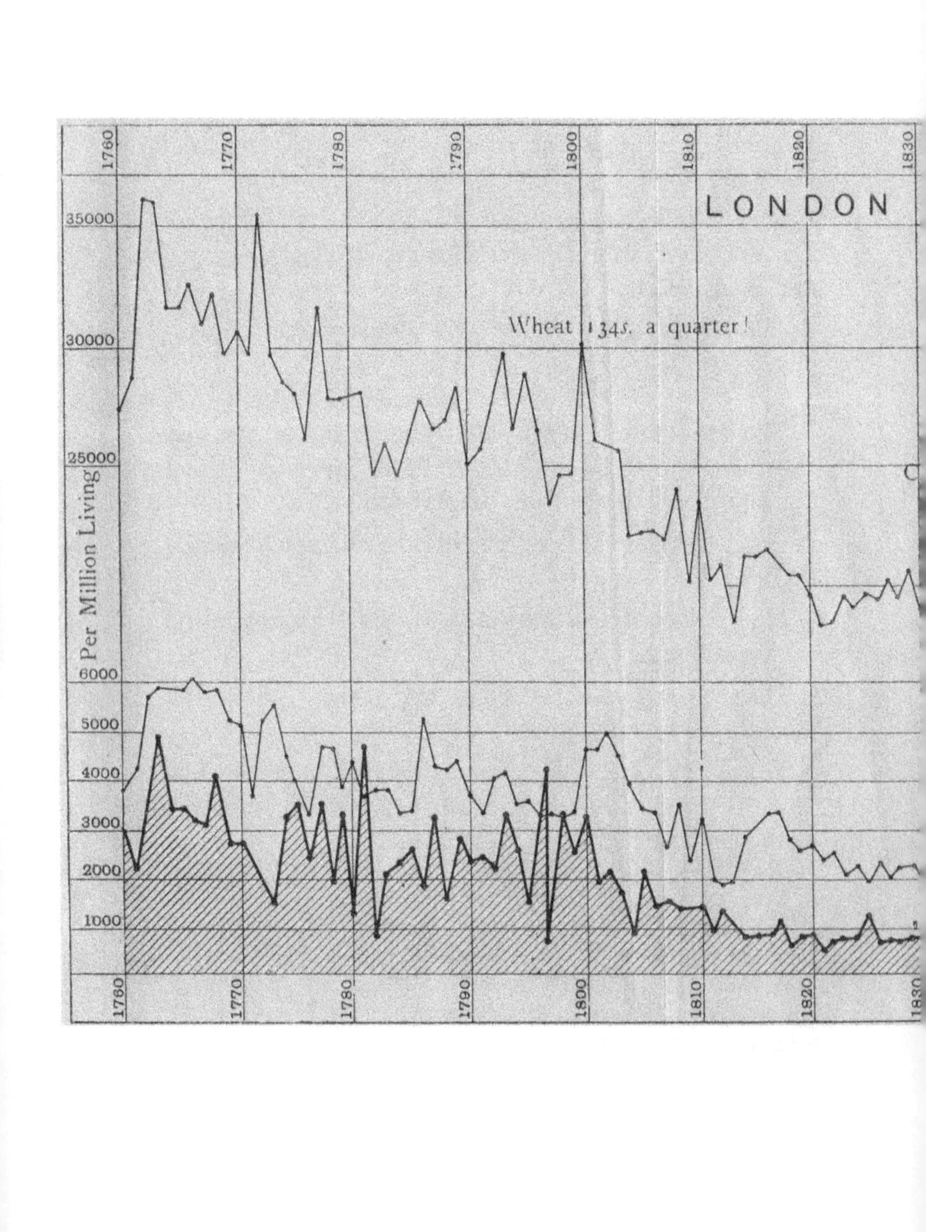
LONDON
Wheat 134s. a quarter!
Per Million Living
35000
30000
25000
6000
5000
4000
3000
2000
1000
1760
1770
1780
1790
1800
1810
1820
1830

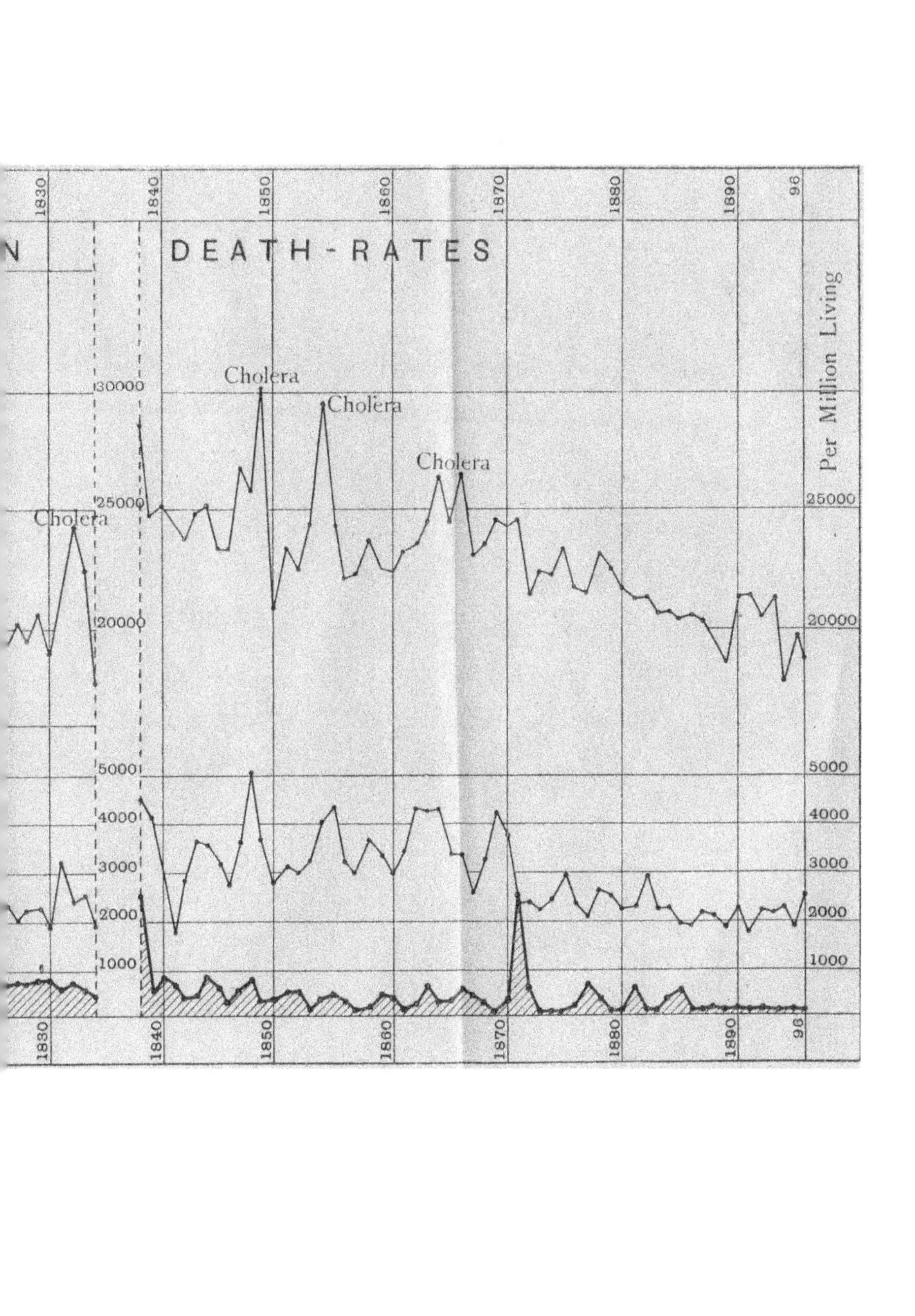

1830
1840
1850
1860
1870
1880
1890
96
N
DEATH-RATES
Per Million Living
Cholera
Cholera
Cholera
Cholera
30000
25000
20000
25000
20000
5000
4000
3000
2000
1000
5000
4000
3000
2000
1000

## DIAGRAM II.

SHOWING DEATH-RATES FROM THE CHIEF ZYMOTIC DISEASES IN LONDON FROM 1838 TO 1896.

From the Registrar-General's Annual Summary, 1896, Table 14, page xxxiii., and 1888, Table 12, for first nine years.

These diagrams show the same facts as Dr. Whitelegge's Diagram E. in the Sixth Report of the Royal Commission, page 660, but in a simpler form.

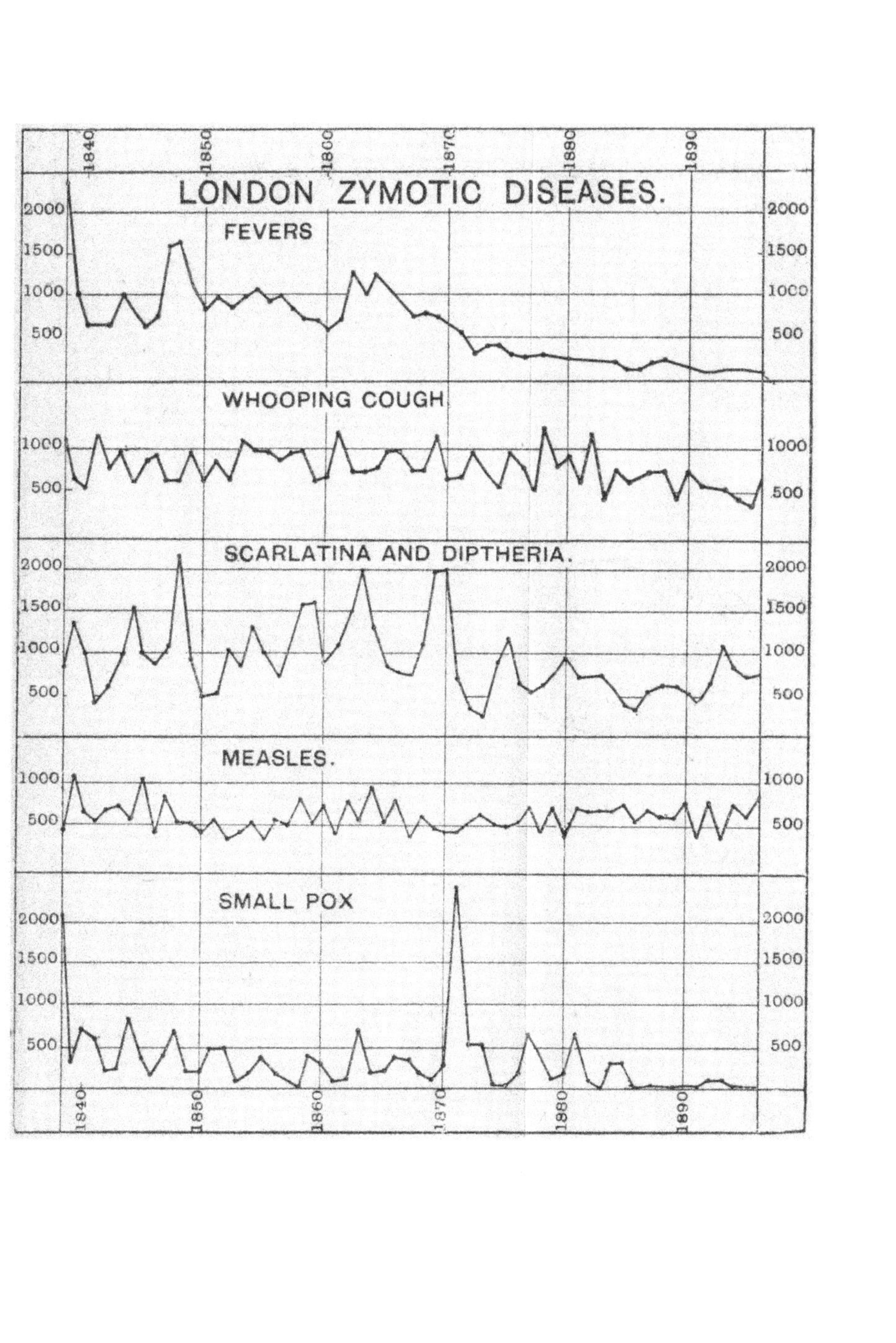
LONDON ZYMOTIC DISEASES.
FEVERS
WHOOPING COUGH
SCARLATINA AND DIPTHERIA.
MEASLES.
SMALL POX
1840
1850
1860
1870
1880
1890
2000
1500
1000
500

## DIAGRAM III.

SMALL-POX, VACCINATIONS, ZYMOTICS, AND TOTAL DEATH-RATE IN ENGLAND AND WALES.

Small-pox from Final Report, Tab. B. p. 155, and Registrar-General's Report, 1895, Table 24.

Vaccinations from Final Report, p. 34.

Zymotic diseases from Registrar-General's Report (1895), Table 24, Columns 3 to 9.

Total Death-rate from Registrar-General's Report, 1895, Table 3.

N.B.—Each of the lines showing Death-rates has its own vertical scale showing the rate per million living, in order to allow of the four separate rates being shown on one diagram so that their corresponding rise or fall may be compared.

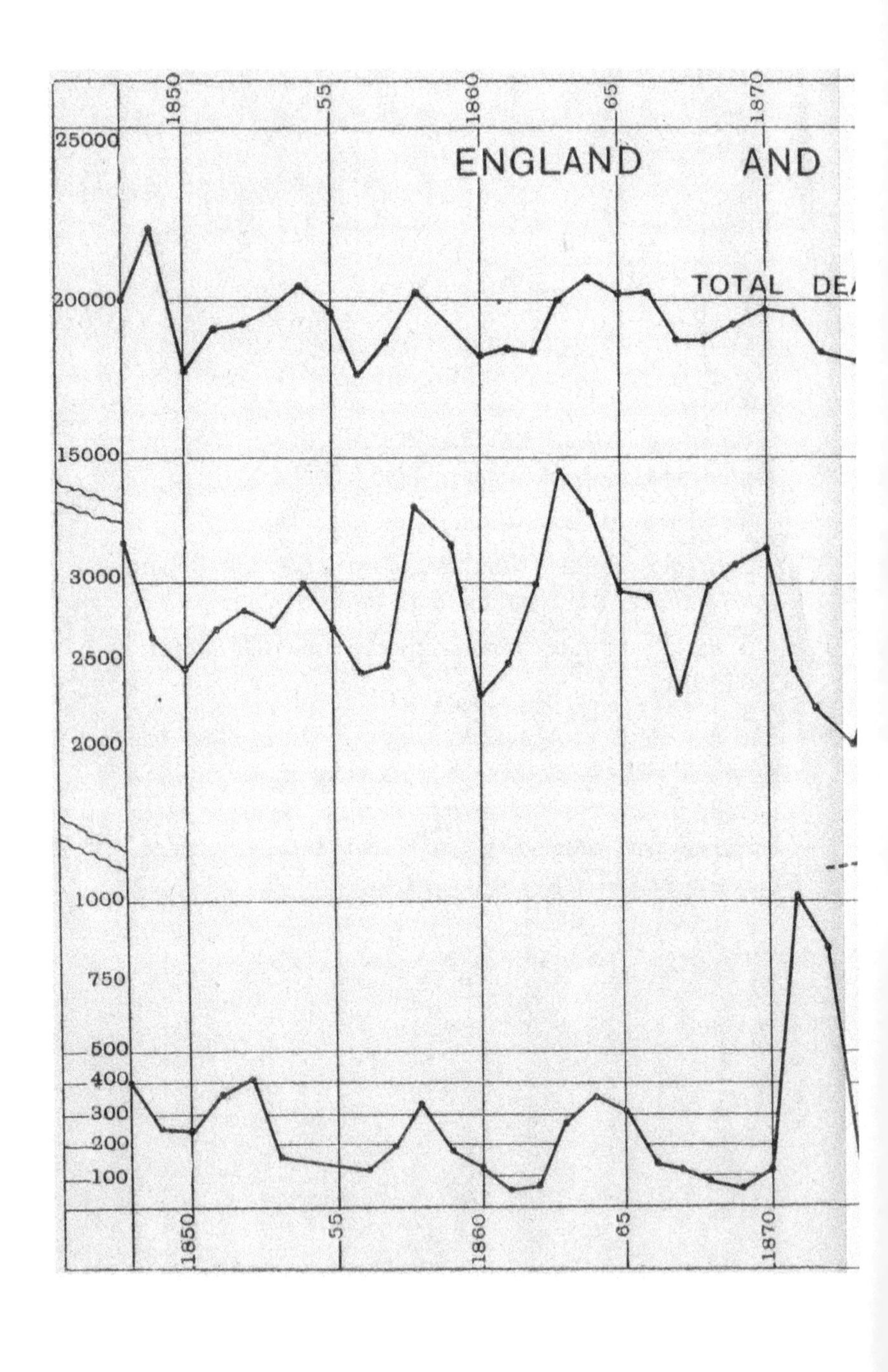

ENGLAND AND
TOTAL DE
25000
20000
15000
3000
2500
2000
1000
750
500
400
300
200
100
1850
55
1860
65
1870

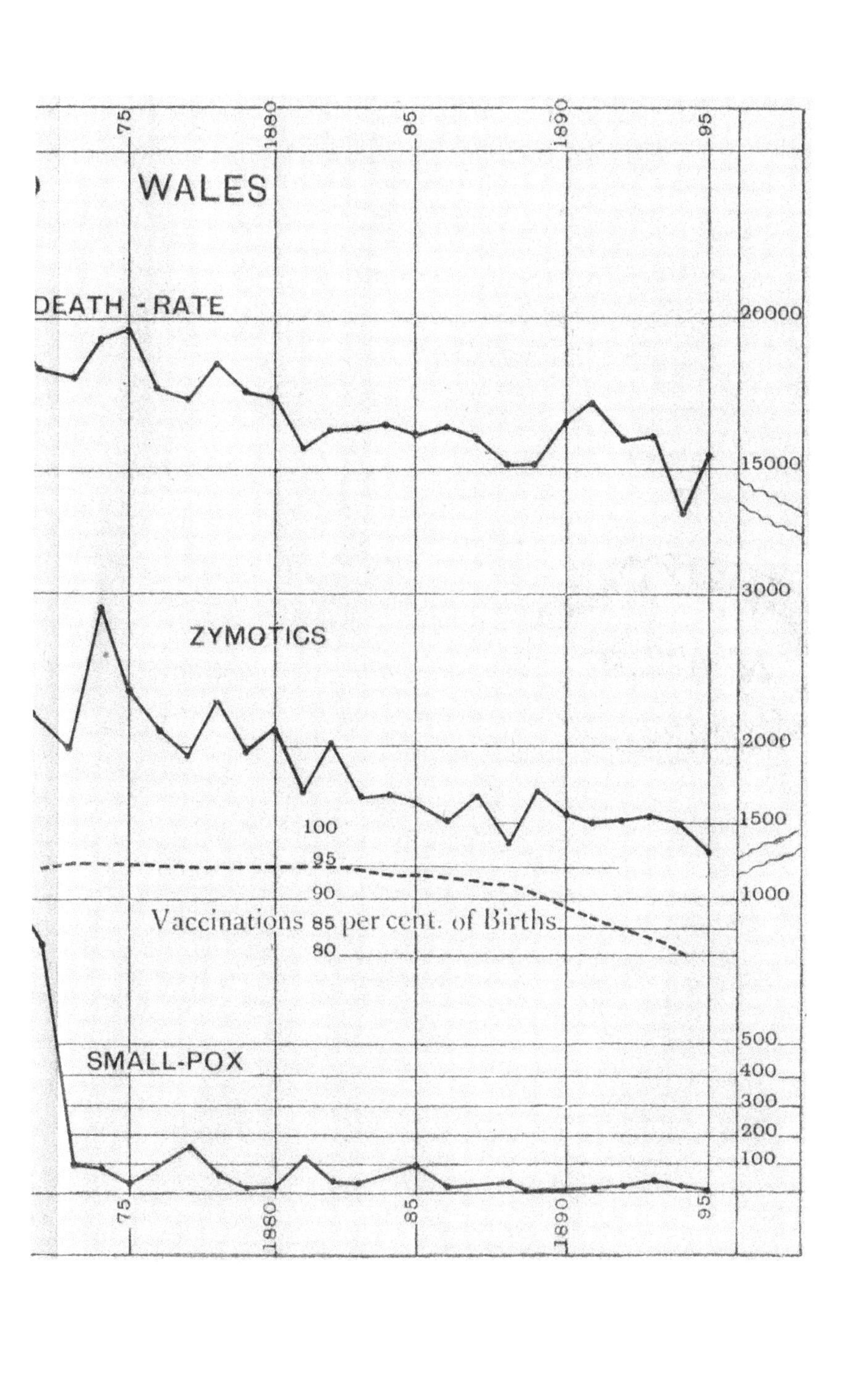

75
1880
85
1890
95
WALES
DEATH - RATE
20000
15000
3000
ZYMOTICS
2000
1500
100
95
90
Vaccinations 85 per cent. of Births
80
1000
500
400
300
200
100
SMALL-POX
75
1880
85
1890
95

## DIAGRAM IV.

Comparison of Scotland and Ireland as regards their Death-rates from Small-pox and two Zymotics (Measles and Scarlet-fever).

From Tables given in the Roy. Comm. Final Report. (See pages 35, 37, 42, and 44.)

Solid lines. Small-pox (shaded for distinctness).

Dotted lines. Two Zymotics.

Both per million living.

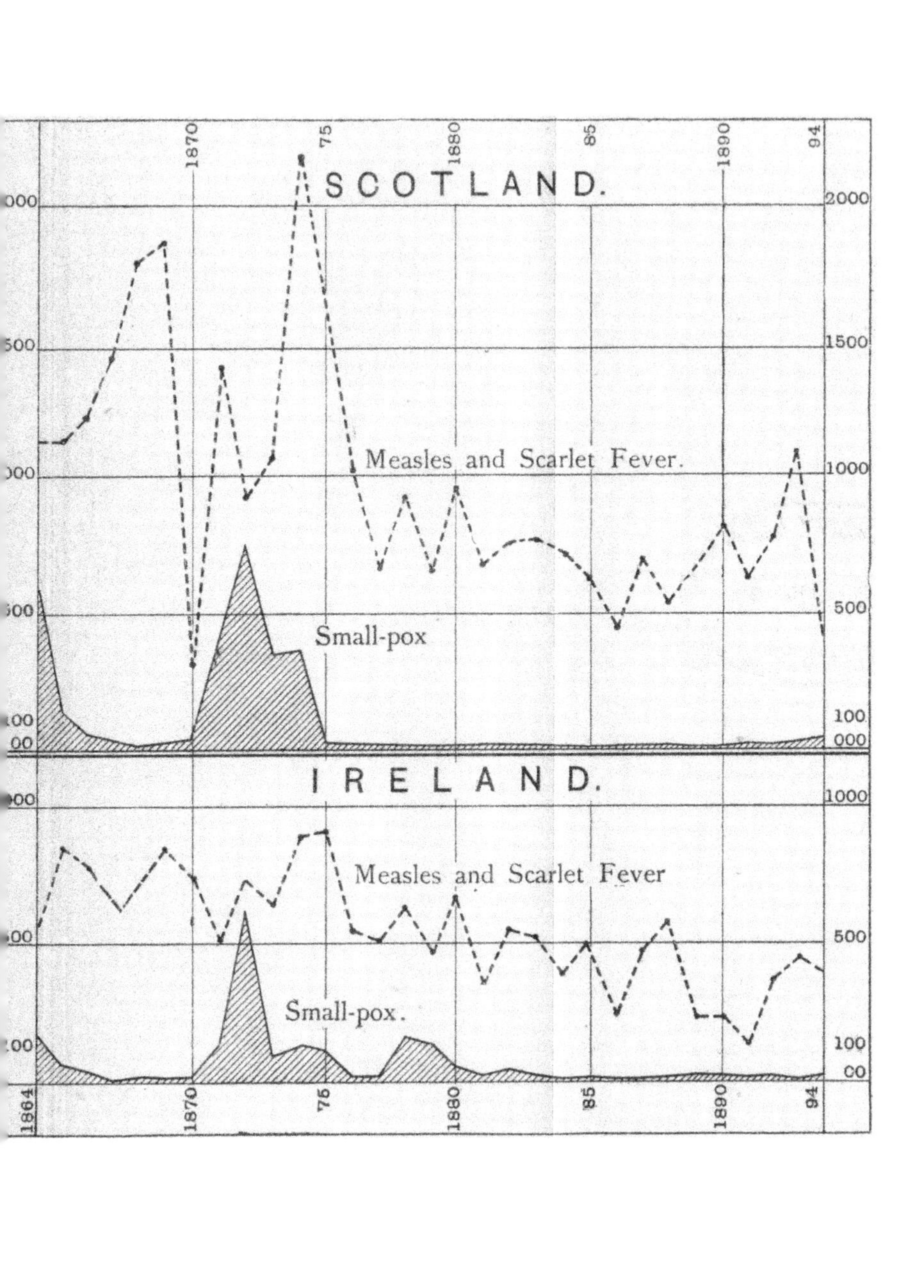
SCOTLAND.
Measles and Scarlet Fever.
Small-pox
IRELAND.
Measles and Scarlet Fever
Small-pox.
1864
1870
75
1880
85
1890
94
2000
1500
1000
500
100
000
1000
500
100
00

## DIAGRAM V.

### SWEDEN. SMALL-POX AND TOTAL DEATH-RATES, AND STOCKHOLM SMALL-POX EPIDEMICS.

These death-rates have been calculated by myself from the official tables of Small-pox and total deaths, and populations in the Sixth Report, pages 752–3.

The portion relating to Small-pox agrees with Diagram D, p. 129, in the Third Report of the Commission, but comes to a later date. The figures for the Stockholm epidemics are not given in the Reports of the Royal Commission except as regards the last and greatest of them. The others are from the same authority as in my former diagram—Dr. Berg, head of the Statistical Department at Stockholm, who supplied them to Dr. Pierce as stated in his *Vital Statistics.*

The Upper line, showing the death-rate from all causes, is from the five-year average mortality, and is on a smaller vertical scale (as shown by the figures at the sides) in order to bring it into the same diagram.

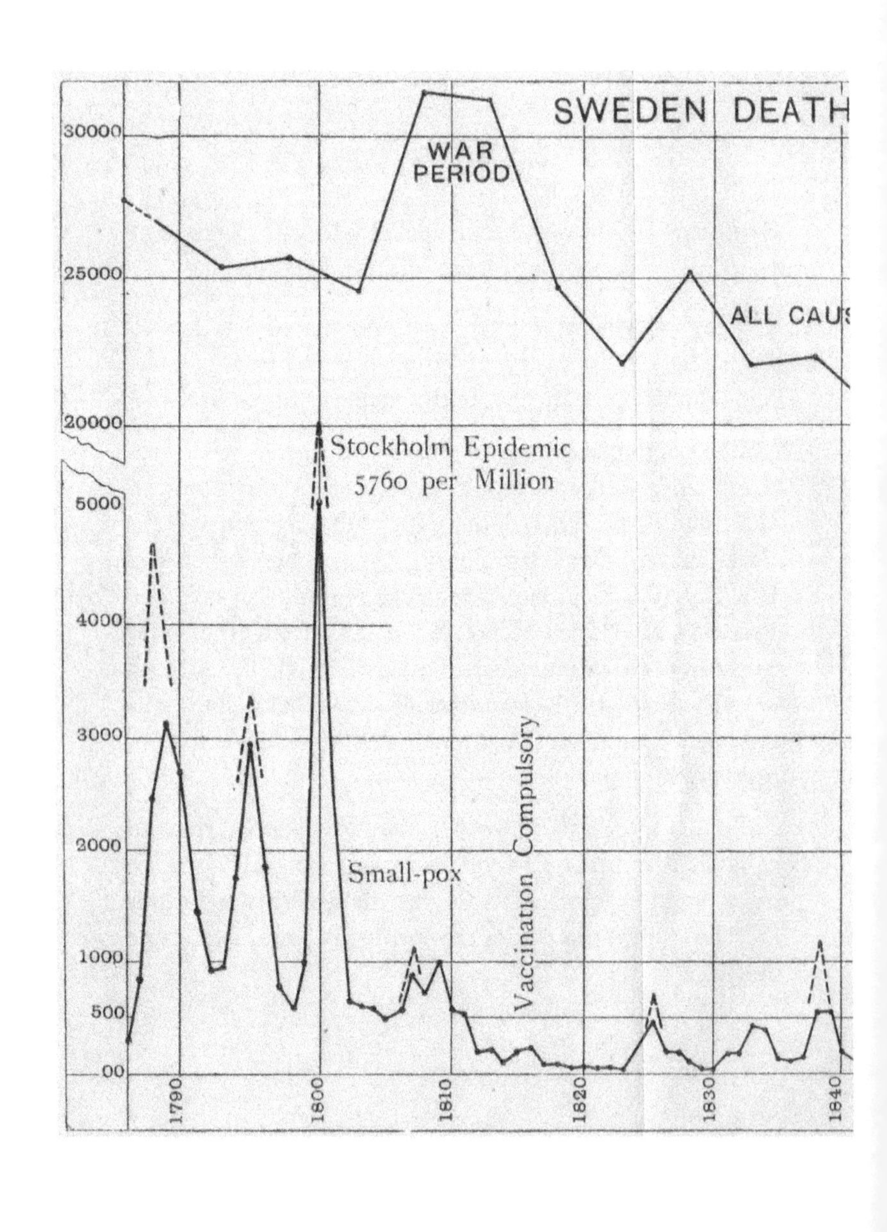
SWEDEN DEATH
WAR PERIOD
ALL CAU
Stockholm Epidemic 5760 per Million
Small-pox
Vaccination Compulsory
30000
25000
20000
5000
4000
3000
2000
1000
500
00
1790
1800
1810
1820
1830
1840

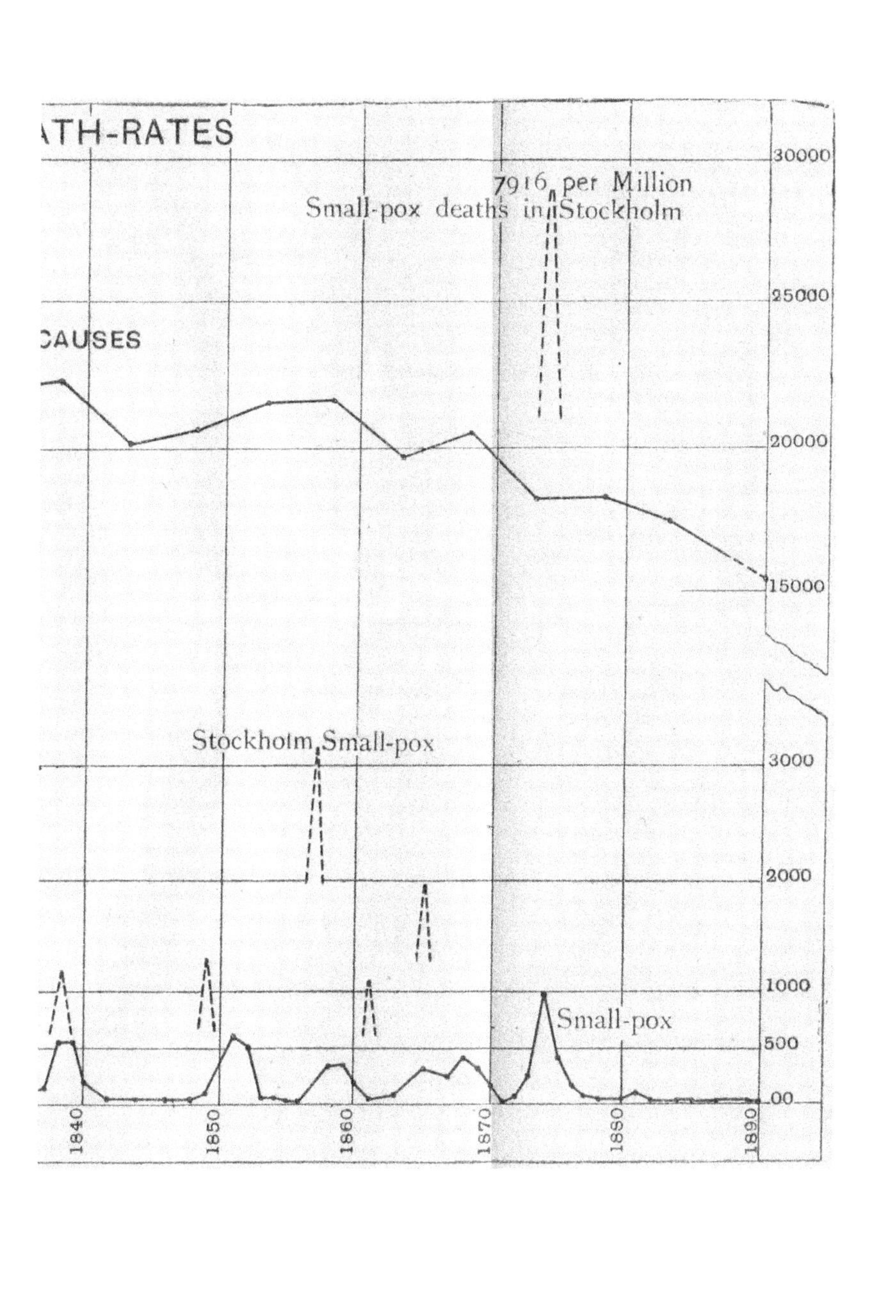

ATH-RATES
CAUSES
7916 per Million
Small-pox deaths in Stockholm
Stockholm Small-pox
Small-pox
30000
25000
20000
15000
3000
2000
1000
500
00
1840
1850
1860
1870
1880
1890

DIAGRAM

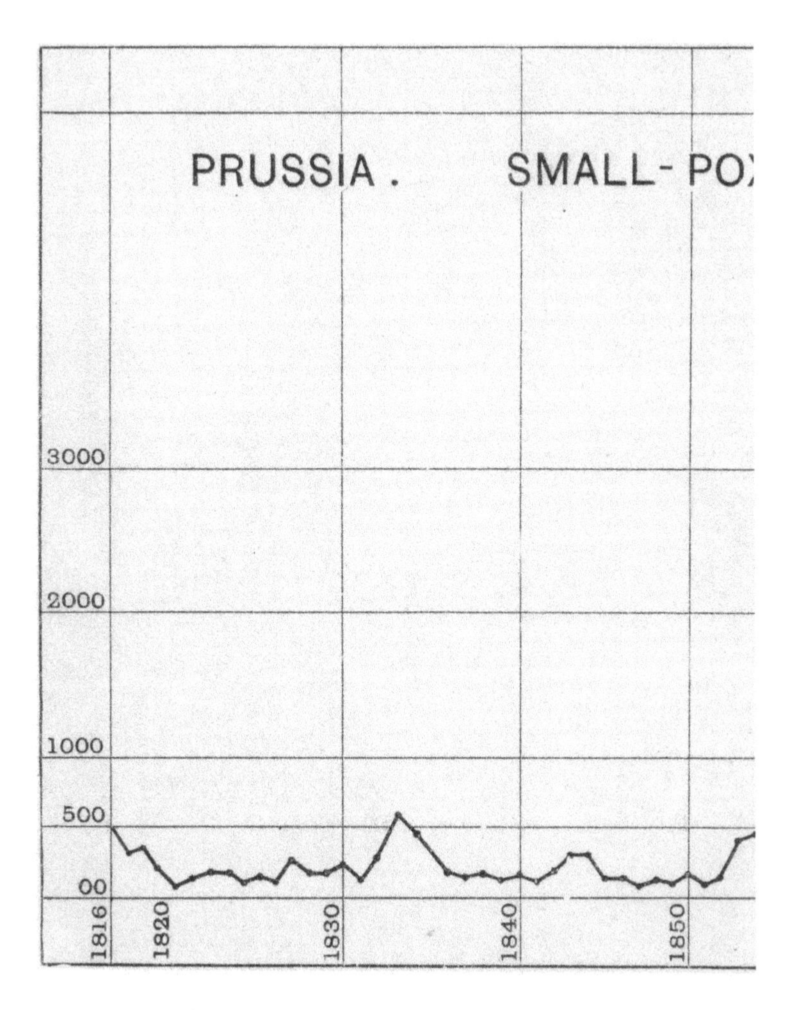


Small-pox death-rates in Prussia ———

From the figures appended to the diagram and the Berlin epidemics from the table at p.231

VI.

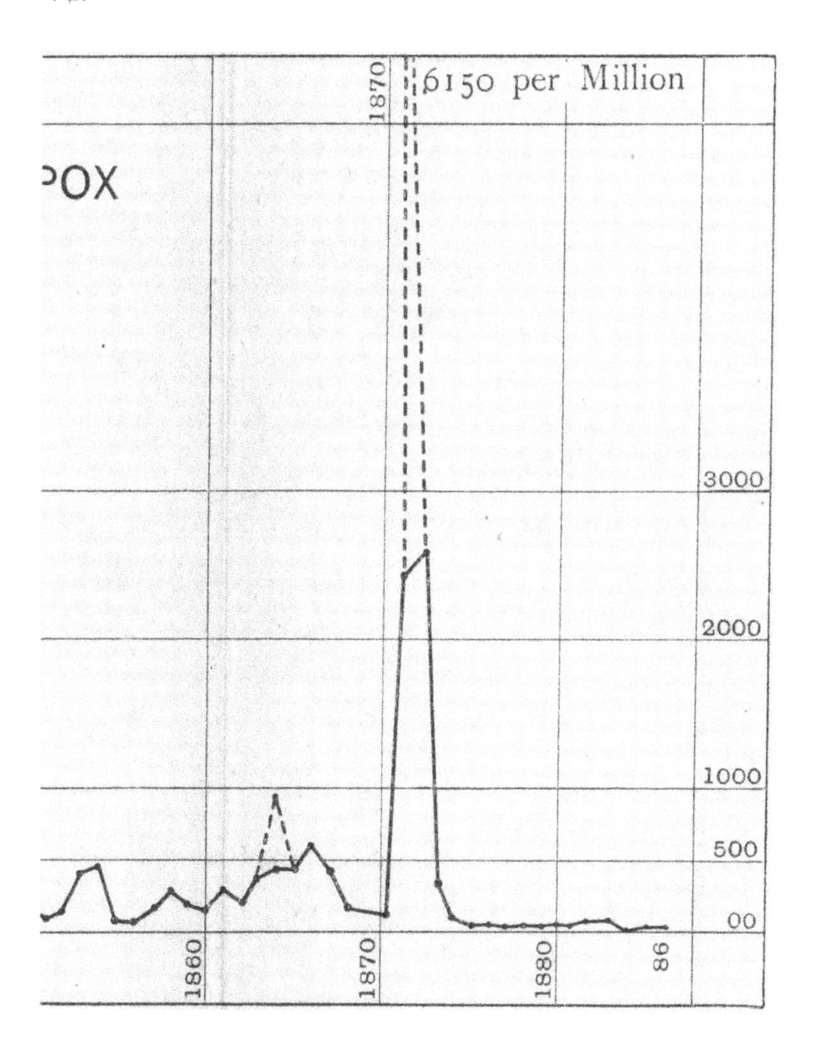


— Epidemics in Berlin -----------

opposite p. 232 of the Second Report,

of the same Report.

## DIAGRAM VII.

BAVARIA. MORTALITY FROM SMALL-POX AND OTHER ZYMOTIC DISEASES IN THE YEARS 1858–73.

From Tables in the Second Report, pp. 337–8.

Bavaria is chosen by Dr. Hopkirk to show the advantages of compulsory vaccination (see Q. 1489, p. 11, and Table facing p. 238, of Second Report).

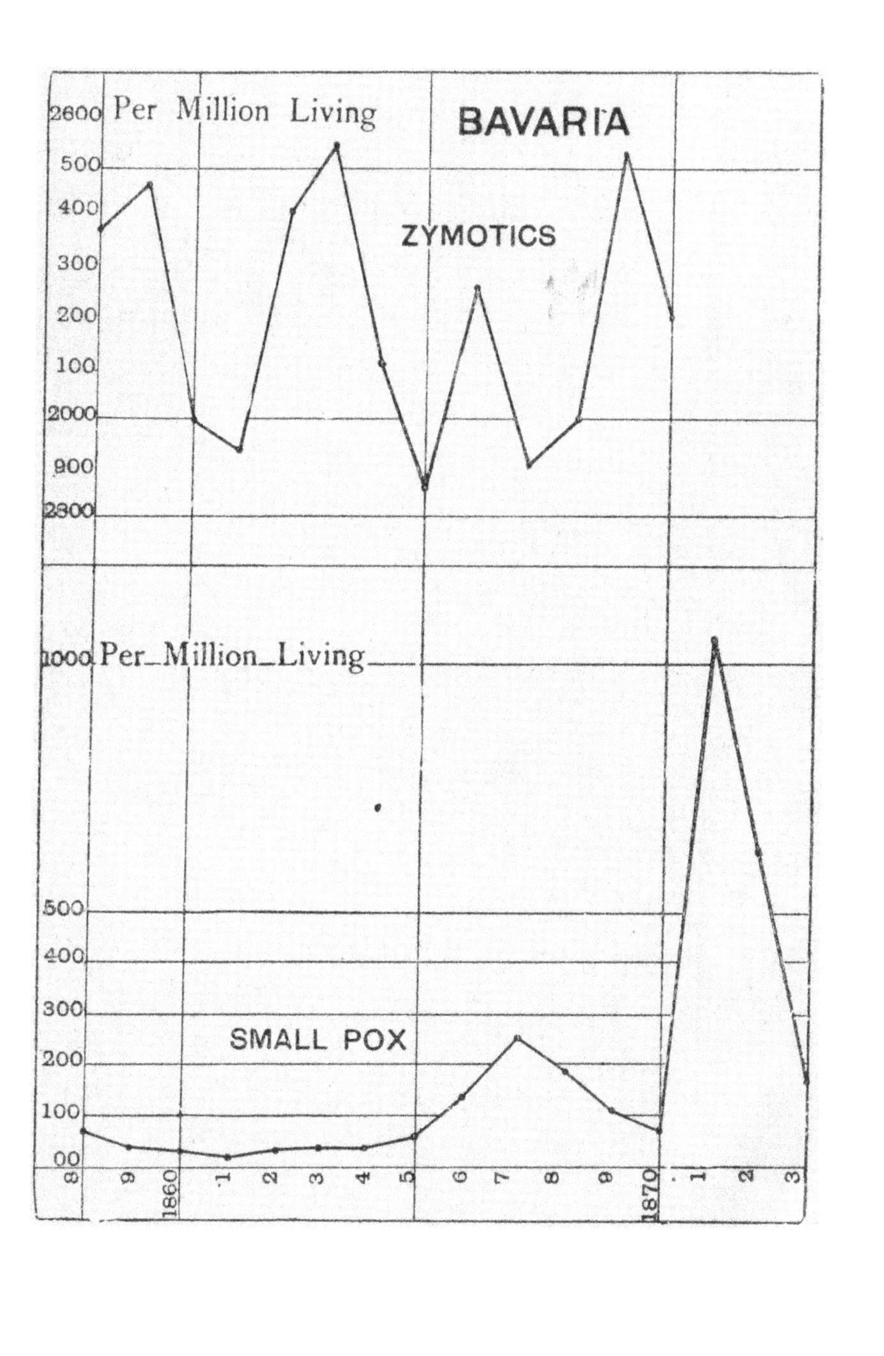

BAVARIA
Per Million Living
2600
500
400
300
200
100
2000
900
2800
ZYMOTICS
1000 Per Million Living
500
400
300
200
100
00
SMALL POX
8
9
1860
1
2
3
4
5
6
7
8
9
1870
1
2
3

# DIAGRAM VIII.

SHOWING THE DEATH-RATES PER MILLION LIVING BY SMALL-POX AND ZYMOTIC DISEASES, FROM 1838 TO 1896, IN LEICESTER.

The dotted line shows the percentage of Vaccinations to Births.

N.B.—Before 1862 private vaccinations have been estimated.

The Upper Thick line shows the death-rate from the following diseases:—Measles, Scarlet Fever, Diphtheria, Typhus, Whooping Cough, Enteric and other Fevers.

The Lower Line, shaded for distinctness, shows the Small-pox death-rate.

Drawn from Mr. Thomas Biggs' Table 19, at p. 440 of the Fourth Report, kindly continued by Mr. Biggs to 1896.

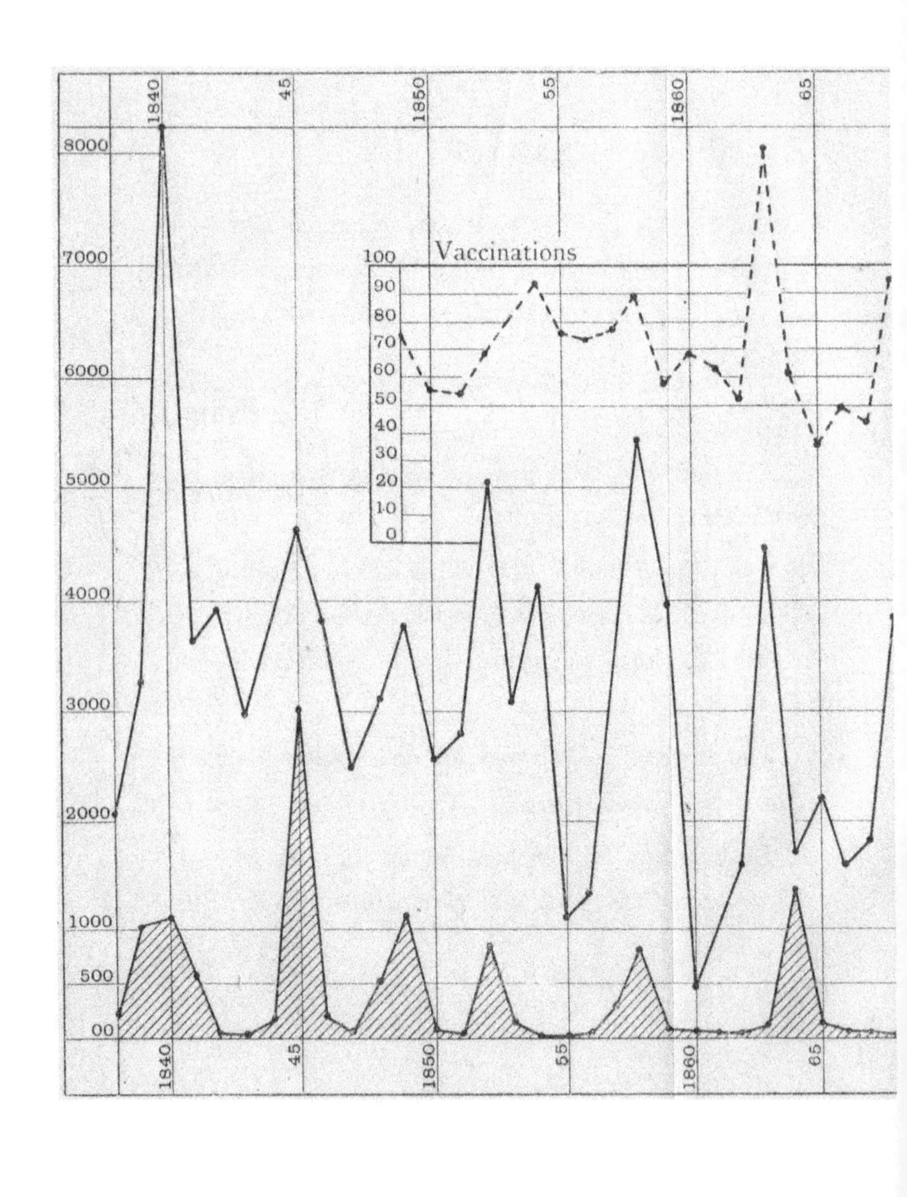
1840
45
1850
55
1860
65
8000
7000
6000
5000
4000
3000
2000
1000
500
00
Vaccinations
100
90
80
70
60
50
40
30
20
10
0
1840
45
1850
55
1860
65

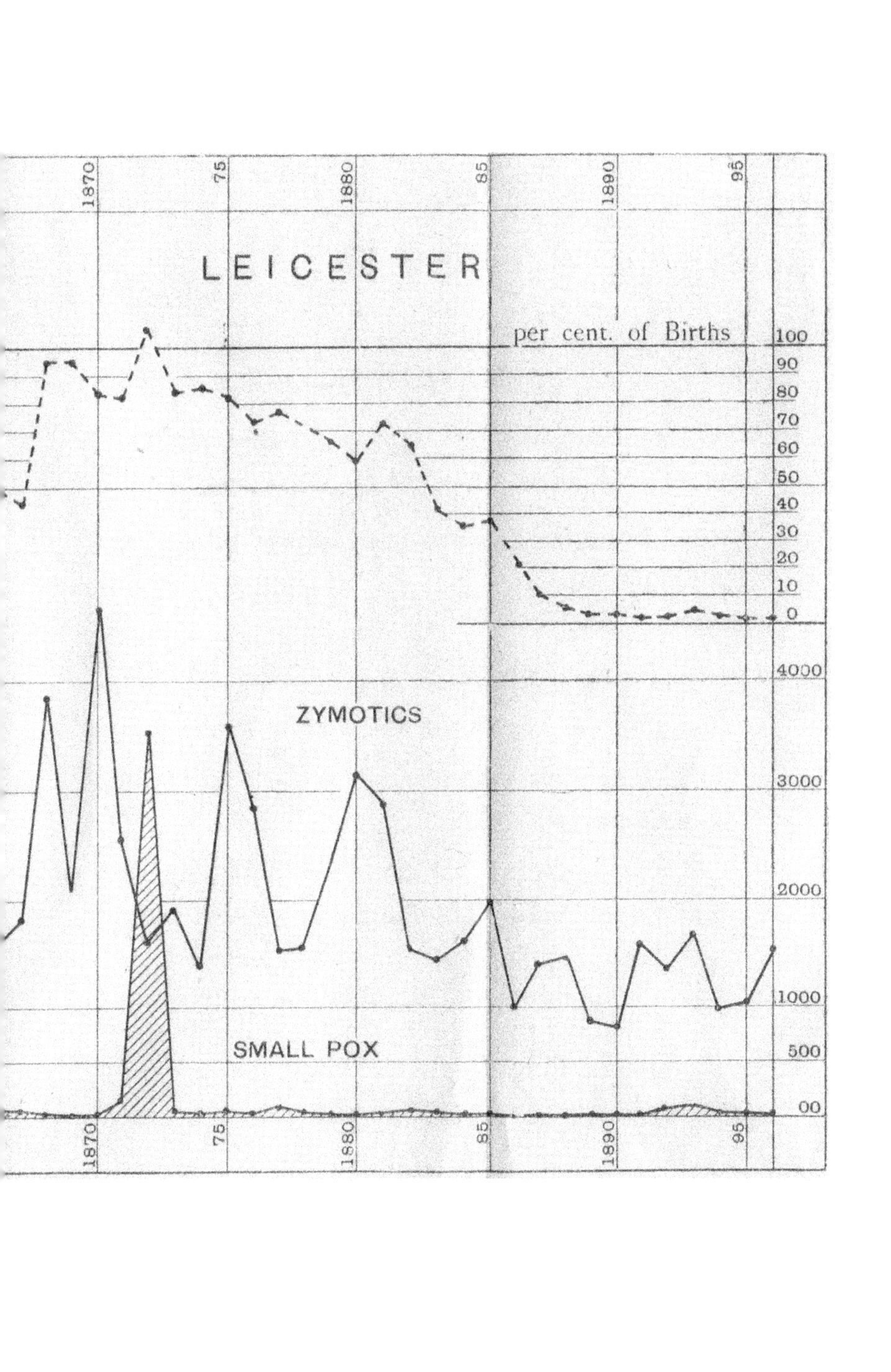

1870
75
1880
85
1890
95
LEICESTER
per cent. of Births
100
90
80
70
60
50
40
30
20
10
0
4000
ZYMOTICS
3000
2000
1000
SMALL POX
500
00
1870
75
1880
85
1890
95

## DIAGRAM IX.

THIS DIAGRAM SHOWS VARIOUS DEATH-RATES IN LEICESTER, IN FIVE-YEAR AVERAGES.

The dotted line shows the percentage of vaccinations to total births.

*Authorities.*

The three Death-rates and the Vaccinations are from Table 34 (p. 450) in the Fourth Report.

The Small-pox death-rate is from Table 45 (p. 461) in same Report.

Figures to continue the diagram to 1896 have been kindly furnished by Mr. Biggs from official sources.

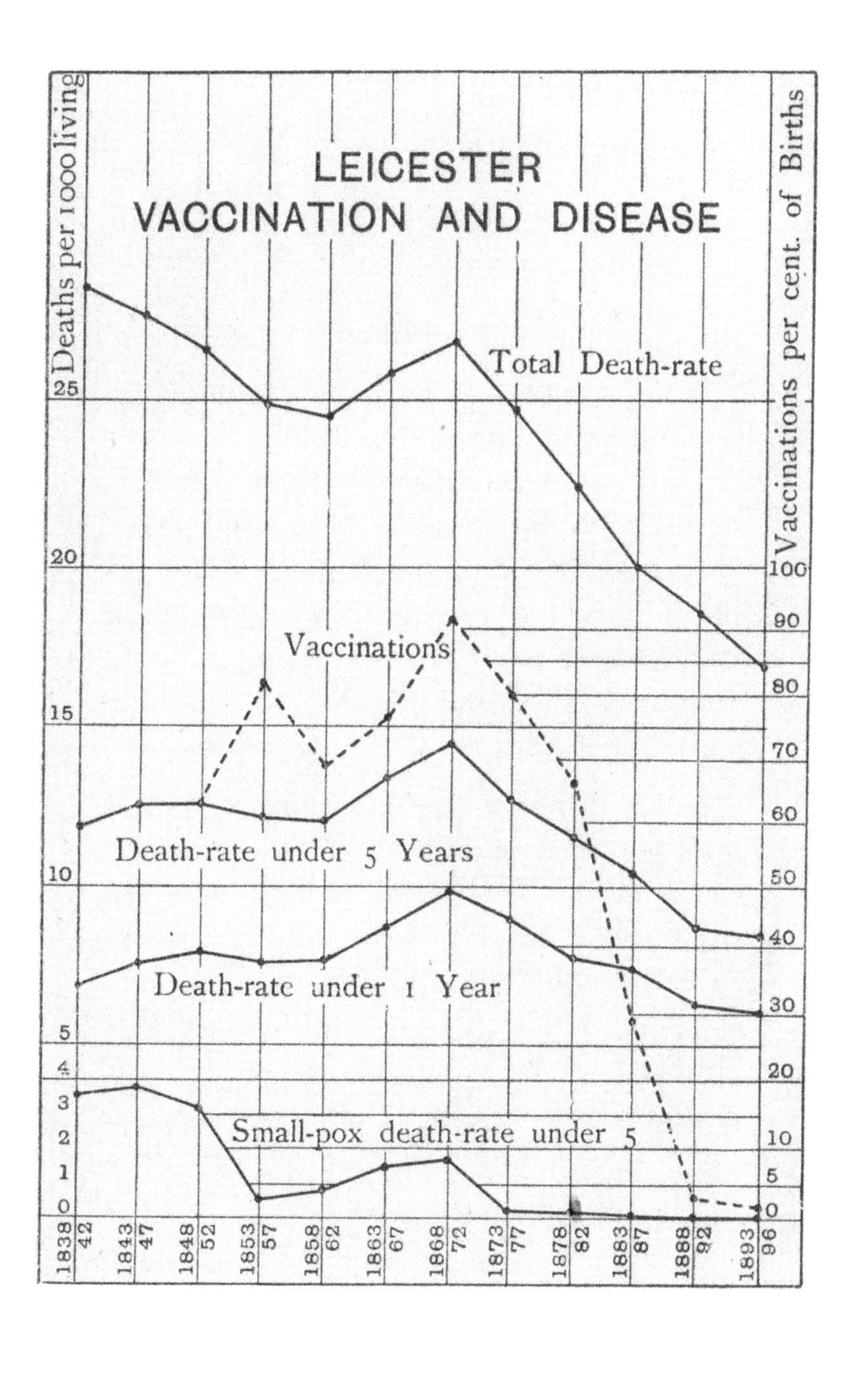
LEICESTER
VACCINATION AND DISEASE
Deaths per 1000 living
Vaccinations per cent. of Births
Total Death-rate
Vaccinations
Death-rate under 5 Years
Death-rate under 1 Year
Small-pox death-rate under 5
25
20
15
10
5
4
3
2
1
0
100
90
80
70
60
50
40
30
20
10
5
0
1838 42
1843 47
1848 52
1853 57
1858 62
1863 67
1868 72
1873 77
1878 82
1883 87
1888 92
1893 96

## DIAGRAM X.

### Infant Mortality.

The upper portion of this diagram shows the Infant Mortality of London from 1730 to 1830, from Dr. Farr's tables in McCulloch's *Statistical Account of the British Empire*, vol. ii., p. 543 (1847). From 1840 to 1890 shows the Infant Mortality of England calculated from the Reports of the Registrar-General (see 3rd Report, p. 197, Table O). Materials for the continuation of Dr. Farr's London Table (under 5 years) are not given by the Registrar-General.

The Lower part of the Table shows, on a larger scale, the Infant Mortality of London, under *one* year, as given by the Registrar-General in his Annual summary for 1891, Table 12, p. xxv., and in his 58th Annual Report, Table 25, p. xci.

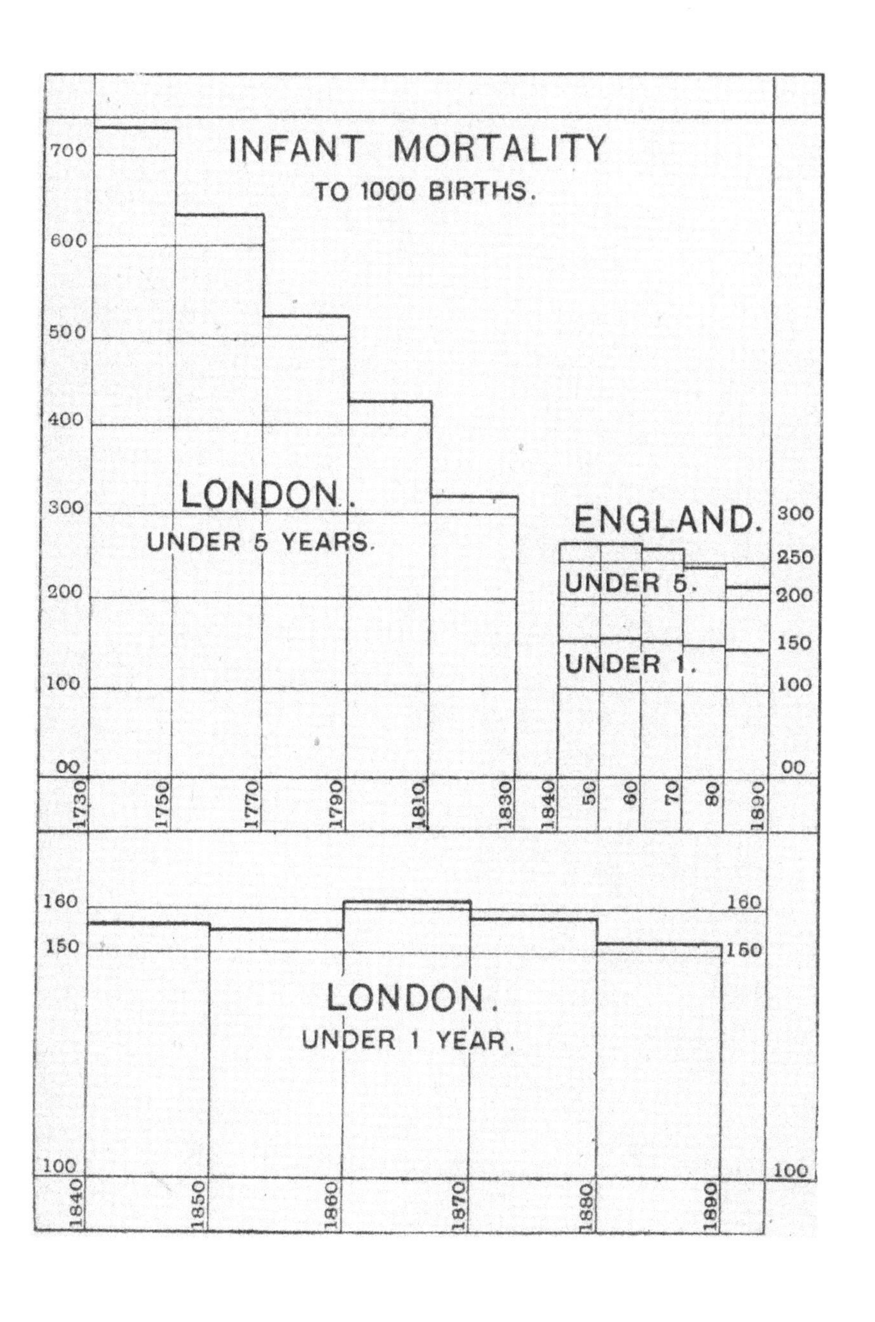
INFANT MORTALITY
TO 1000 BIRTHS.
700
600
500
400
300
200
100
00
LONDON.
UNDER 5 YEARS.
ENGLAND.
UNDER 5.
UNDER 1.
300
250
200
150
100
00
1730
1750
1770
1790
1810
1830
1840
50
60
70
80
1890
160
150
LONDON.
UNDER 1 YEAR.
100
160
150
100
1840
1850
1860
1870
1880
1890

## DIAGRAM XI

### Army and Navy.

Lower Thick line shows the Small-pox mortality per 100,000 in the Army.

Upper Thick line shows the total Disease Mortality in the Army (Home Force).

The two Thin lines show the corresponding Mortalities in the Navy.

*Authorities.*

Total Disease Mortalities, from the Registrar-General's 51st Report, Table 29, and 58th Report, Table 33, for the Army. From Table at p. 254 of Second Report of Roy. Comm. for the Navy.

Small-pox Mortalities from the "Final Report," pp. 86–88.

N.B.—The higher figures (hundreds) show the Disease mortality; the lower figures (tens) show the Small-pox mortality; both per 100,000.

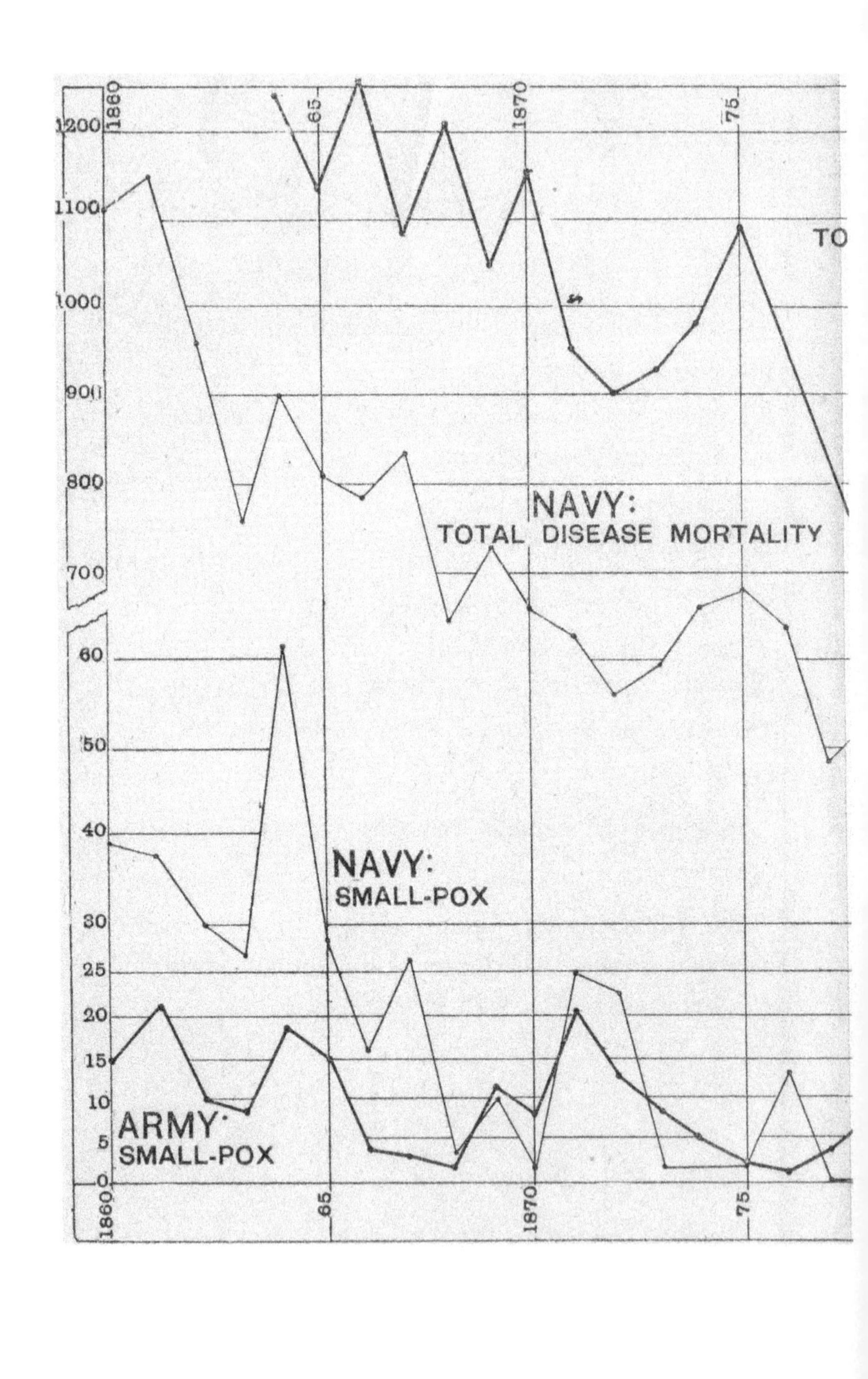

1860
65
1870
75
1200
1100
1000
900
800
700
60
50
40
30
25
20
15
10
5
0
TO
NAVY:
TOTAL DISEASE MORTALITY
NAVY:
SMALL-POX
ARMY:
SMALL-POX
1860
65
1870
75

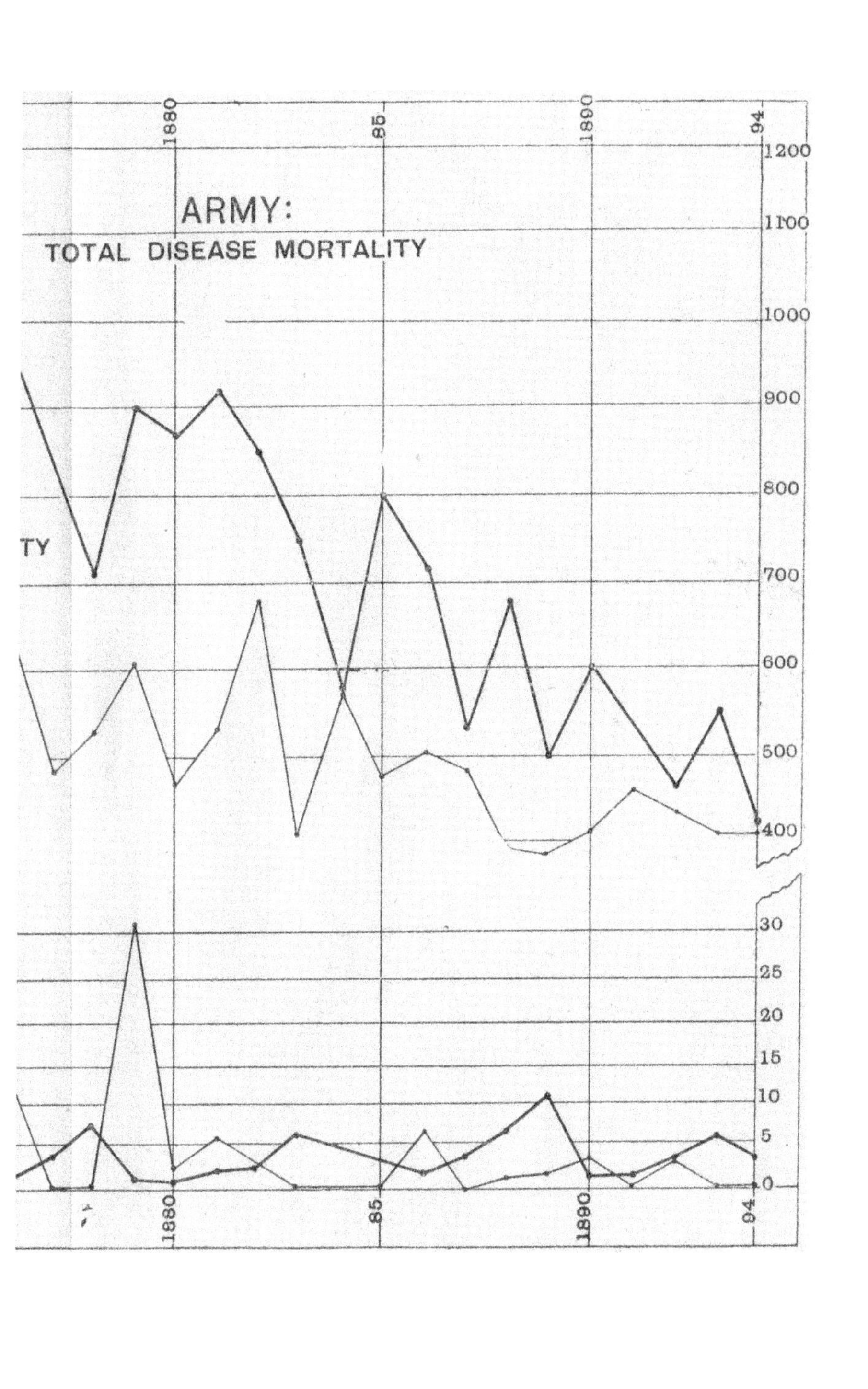

ARMY:
TOTAL DISEASE MORTALITY
TY
1880
85
1890
94
1200
1100
1000
900
800
700
600
500
400
30
25
20
15
10
5
0

## DIAGRAM XII.

Small-pox Mortality per 100,000.

The Army and Navy as compared with Ireland.

From the earliest year given for Ireland in the Reports of the Royal Commission.

*Authorities.*

Army, 2nd Report, Table C., p. 278.

Navy, 2nd Report, Table C., p. 254.

Both supplemented for the last six years by the Final Report," pp. 86–88.

Ireland. Table on p. 57 of "Final Report" corrected to ages 15–45 by adding one-tenth according to the Table J. at p. 274 of 2nd Report.

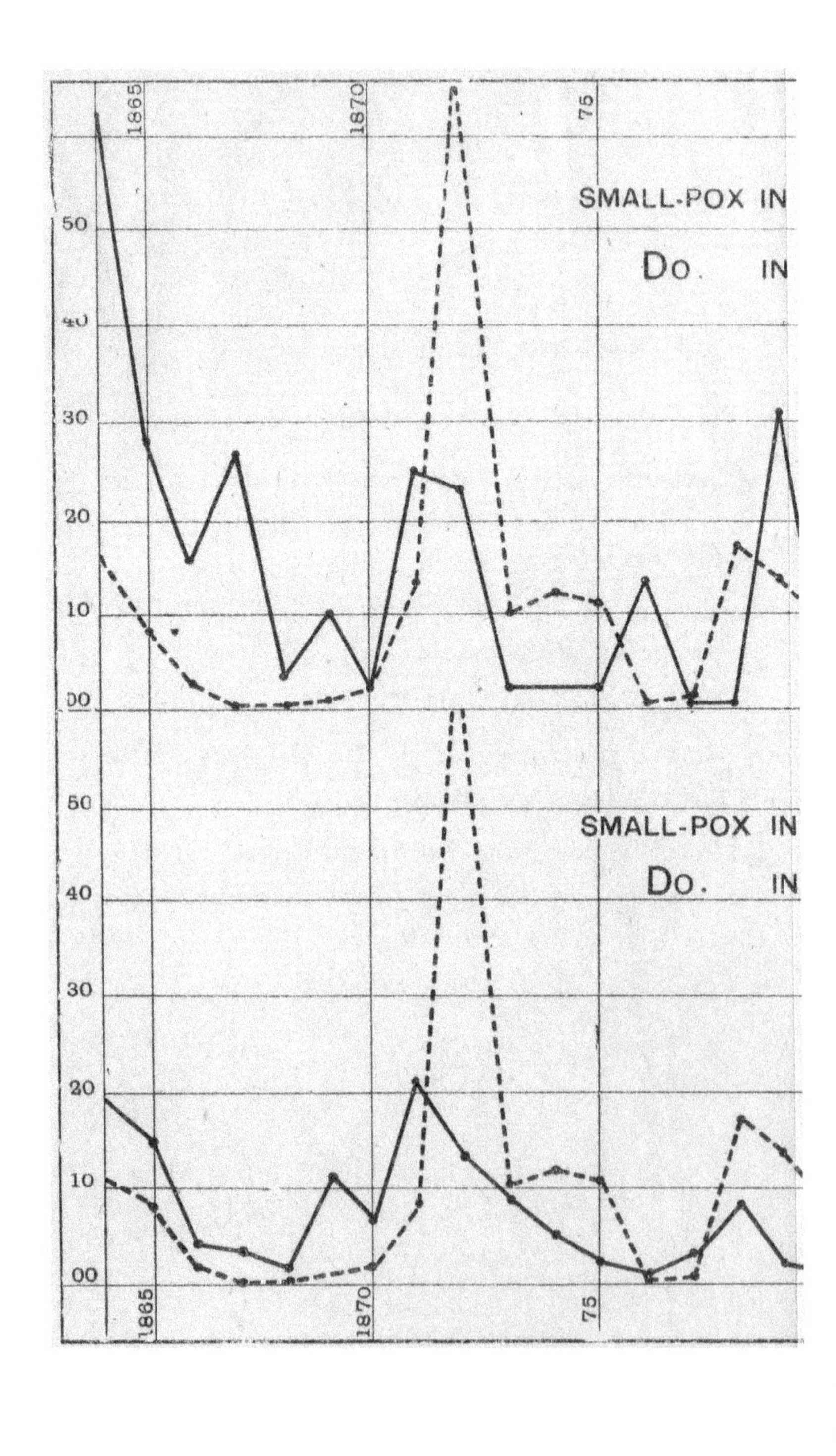

1865
1870
75
SMALL-POX IN
Do. IN
50
40
30
20
10
00
SMALL-POX IN
Do. IN
50
40
30
20
10
00
1865
1870
75

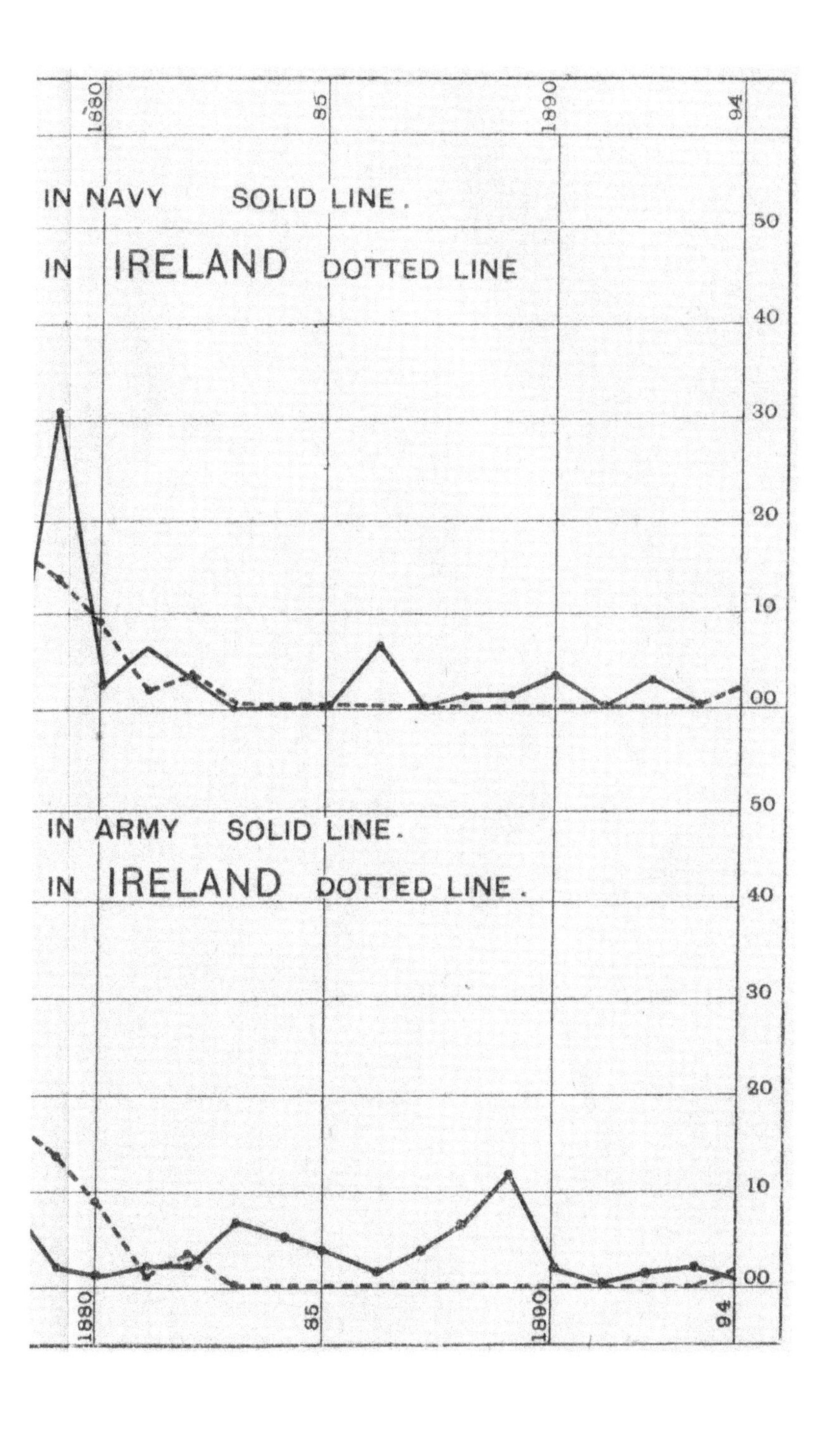

1880
85
1890
94
IN NAVY SOLID LINE.
IN IRELAND DOTTED LINE
50
40
30
20
10
00
IN ARMY SOLID LINE.
IN IRELAND DOTTED LINE.
50
40
30
20
10
00
1880
85
1890
94

For EU product safety concerns, contact us at Calle de José Abascal, 56–1°, 28003 Madrid, Spain or eugpsr@cambridge.org.

www.ingramcontent.com/pod-product-compliance
Ingram Content Group UK Ltd.
Pitfield, Milton Keynes, MK11 3LW, UK
UKHW010353140625
459647UK00010B/1020